FOURIER OPTICS: An Introduction

ELLIS HORWOOD SERIES IN PHYSICS

Series Editor: **C. Grey Morgan**, Department of Physics, University College, Swansea

ELECTRONICS FOR THE PHYSICIST WITH APPLICATIONS
C. F. G. DELANEY, Trinity College, University of Dublin

MAN-MADE GEMSTONES
D. ELWELL, Center for Materials Research, Stanford University, California

PHYSICS FOR ENGINEERS AND SCIENTISTS
D. ELWELL, Stanford University, California, and A. J. POINTON, Portsmouth Polytechnic

VIBRATION AND WAVES
W. GOUGH, J. P. G. RICHARDS and R. P. WILLIAMS, University College, Cardiff

AN INTRODUCTION TO STATISTICAL PHYSICS
W. G. V. ROSSER, University of Exeter

PARTICLE DETECTION TECHNIQUES
M. C. SCOTT, University of Birmingham

FOURIER OPTICS: An Introduction
E. G. STEWARD, The City University, London

SUPERFLUIDITY AND SUPERCONDUCTIVITY 2nd Edition
D. R. TILLEY and J. TILLEY, University of Essex

RHEOLOGICAL TECHNIQUES
R. W. WHORLOW, University of Surrey

FOURIER OPTICS:
An Introduction

E. G. STEWARD, D.Sc., F.Inst. P.
Professor of Physics and Molecular Medicine
The City University, London

ELLIS HORWOOD LIMITED
Publishers · Chichester

Halsted Press: a division of
JOHN WILEY & SONS
New York · Brisbane · Chichester · Ontario

First published in 1983 by
ELLIS HORWOOD LIMITED
Market Cross House, Cooper Street, Chichester, West Sussex, PO19 1EB, England

The publisher's colophon is reproduced from James Gillison's drawing of the ancient Market Cross, Chichester.

Distributors:

Australia, New Zealand, South-east Asia:
Jacaranda-Wiley Ltd., Jacaranda Press,
JOHN WILEY & SONS INC.,
G.P.O. Box 859, Brisbane, Queensland 40001, Australia

Canada:
JOHN WILEY & SONS CANADA LIMITED
22 Worcester Road, Rexdale, Ontario, Canada.

Europe, Africa:
JOHN WILEY & SONS LIMITED
Baffins Lane, Chichester, West Sussex, England.

North and South America and the rest of the world:
Halsted Press: a division of
JOHN WILEY & SONS
605 Third Avenue, New York, N.Y. 10016, U.S.A.

© 1983 E. G. Steward/Ellis Horwood Limited.

British Library Cataloguing in Publication Data
Steward, E.G.
Fourier optics. – (Ellis Horwood Physics series)
1. Optics, Physical
I. Title
535'.2   QC395.2

Library of Congress Card No. 83-9052

ISBN 0-85312-631-3 (Ellis Horwood Ltd., Publishers – Library Edn.)
ISBN 0-85312-632-1 (Ellis Horwood Ltd., Publishers – Student Edn.)
ISBN 0-470-27454-9 (Halsted Press)

Typeset in Press Roman by Ellis Horwood Limited.
Printed in Great Britain by R. J. Acford, Chichester.

COPYRIGHT NOTICE –
All Rights Reserved. No part of this publication may be reproduced, stored in a retrieval system, or transmitted, in any form or by any means, electronic, mechanical, photocopying, recording or otherwise, without the permission of Ellis Horwood Limited, Market Cross House, Cooper Street, Chichester, West Sussex, England.

Table of Contents

Author's Preface 9

1 Preliminaries
1.1 Introduction. 11
1.2 Coherence and light sources. 16
    1.2.1 Temporal coherence. 16
    1.2.2 Spatial coherence 17
1.3 Optical image formation 19
    1.3.1 Fraunhofer diffraction 20
    1.3.2 Lens aperture 25
1.4 Interference by division of amplitude 25

2 Fraunhofer diffraction
2.1 Introduction. 28
2.2 Single-slit pattern 28
2.3 Circular aperture 32
2.4 Double aperture 36
    2.4.1 Two slits 36
    2.4.2 Two circular apertures 39
2.5 N-slit grating. 39
2.6 2-dimensional gratings 43
2.7 Crystals as 3-dimensional gratings 44

3 Fourier series and periodic structures
3.1 Introduction. 49
3.2 Fourier series 50
3.3 Determining Fourier coefficients: even functions. 52
3.4 Optical and crystal diffraction gratings: physical interpretation
        of Fourier terms. 54
    3.4.1 Optical diffraction. 54
    3.4.2 Crystal diffraction. 55

3.5	Fourier series: general formulations		56
	3.5.1	The sine and cosine series	57
	3.5.2	Exponential notation	57
	3.5.3	Space and time	60

4 Fourier transforms, convolution and correlation

4.1	Introduction		61
4.2	The Fourier transform and single-slit diffraction		61
4.3	The grating pattern as a product of transforms		67
4.4	Convolution		70
	4.4.1	Introduction	70
	4.4.2	The grating as a convolution	73
4.5	The convolution theorem and diffraction		74
4.6	Fourier transforms and light waves		75
4.7	Correlation		77
	4.7.1	Autocorrelation theorem (Wiener–Khinchin theorem)	81

5 Optical imaging and processing

5.1	Introduction		83
5.2	Incoherent optical imaging		86
	5.2.1	Determination of transfer functions	88
5.3	Coherent optical imaging		90
	5.3.1	Periodic objects	90
	5.3.2	Non-periodic objects	93
	5.3.3	Illustrations: optical transforms	94
5.4	Holography		102
5.5	Optical processing		106
	5.5.1	Coherent processing	106
	5.5.2	Incoherent processing	115

6 Interferometry and radiation sources

6.1	Introduction		119
6.2	Michelson's stellar interferometer		120
	6.2.1	Introduction	120
	6.2.2	Fringe visibility aspects	123
6.3	Michelson's spectral interferometer		127
	6.3.1	Introduction	127
	6.3.2	Fringe visibility and spectral distribution	131
6.4	Partial coherence, correlation, and visibility		133
	6.4.1	Spatial coherence	137
	6.4.2	Temporal coherence	138
6.5	Fourier transform spectroscopy		139

6.6	Applications in astronomy	145
	6.6.1 Aperture synthesis	146
	6.6.2 The intensity interferometer	153

Appendices
A The scalar-wave description of electromagnetic waves ... 157
B The Stokes treatment of reflection and refraction ... 163
C Diffraction of X-rays by crystals. The equivalence of the Laue conditions and the Bragg reflection concept ... 164
D The electromagnetic spectrum. Approximate ranges of principal, named regions ... 169
E Useful formulae ... 170

References ... 171

Bibliography ... 175

Index ... 178

Author's Preface

This book is intended for students who seek a simple introduction to the Fourier principles of modern optics and an insight into the similar role they play in other branches of science and engineering. Fourier transforms, associated with the operations of convolution and correlation, form the basis not only of image formation and processing with lens systems, but also of studies ranging from the atomic structure of matter to the galactic structure of the universe, and some modern methods of spectroscopy. They also apply to the communication and information sciences of electrical engineering, including the processing of information that is not optical in origin. As a supplement to the undergraduate mainstream textbooks on optics the book bridges the gap to the more advanced treatises in the various specialized fields.

Incoherent imaging is described from the point of view of convolution and transfer functions, with emphasis on the features of linearity and invariance shared with many types of electrical network (non-linear systems being outside the scope of this book). The double Fourier transformation process of coherent image formation is illustrated with particular reference to its application in X-ray crystallography.

As an introduction to the wide range of applications of optical filtering and image processing the basic ideas of amplitude-, phase- and holographic filtering are described, with illustrations drawn from optical and electron microscopy, and from the developing field of pattern recognition. Energy-spectrum correlation and geometrical optics-based processing are also briefly described.

Historical aspects are mentioned wherever they particularly help to convey the spirit of science and show how developments take place: to understand fully the present state of a subject, and where it is going, it is essential to know where it has come from. For example, in Chapter 6 it is explained how the modern uses of interferometry in radio and optical astronomy, and spectroscopy, have their origins in Michelson's stellar and spectral interferometers: the methods that Michelson devised, and his brilliant awareness of their potential, form the natural starting point for understanding their present-day counterparts.

To encourage and help the reader to consult the whole span of the literature the text is referenced throughout, in addition to the inclusion of a bibliography.

I have assumed that the reader already has a knowledge of mathematics up to the usual university entrance level, all additional mathematics being developed in the context in which it is used. In the main, the treatment is restricted to one dimension so as to keep the mathematical statements as visually digestible as possible.

It is a great pleasure to record my sincere thanks to Dr R. S. Longhurst for stimulating discussions and suggestions at many stages in the preparation of the final manuscript: to be able to tap his experience as author of a major undergraduate textbook on optics has been invaluable. Similarly, I am grateful to Professor Henry Lipson who kindly read and commented on Chapters 1, 2 and 5, and who has been a source of encouragement to me over many years; and to Professor Sir Martin Ryle for helpful comments on the section on astronomy in Chapter 6. My thanks also go to my colleague Professor M. A. Jaswon (Head of Mathematics Department at The City University) for helpful comments on my presentation of some of the mathematics.

I am grateful to Roger Meggs (Physics Department) for assistance in the preparation of the figures, and to Peter Clover (Audiovisual Aids) for making all the ink-drawings.

February 1983   E. G. Steward

1

Preliminaries

1.1 INTRODUCTION

This book is primarily concerned with the Fourier aspects of two inter-related topics in physical optics: (i) the formation and processing of images and (ii) the study of the spatial distribution and spectral composition of radiation sources. They are brought together within the same covers because they are inter-related in a number of ways, and when such relationships exist there is, usually, as here, an increased benefit to be obtained from studying the topics together. Their wider implications, referred to in the Preface, will be introduced as they arise.

As we work through these pages we can think of both topics in the context of visible light, but we also deal with some important applications in other parts of the electromagnetic spectrum. In image formation this takes us from the optical region to the way in which X-rays are used to deduce the atomic structure of matter, and at the other extreme, to astronomy and the structure of the universe. In spectroscopy, Fourier-based methods are now used over a wide spectral range.

The mathematical methods named after **J. B. J. Fourier** are extremely powerful in these topics. They are introduced in Chapters 3 and 4, and are used extensively in the closing chapters. For the mathematical representation of light the **'scalar-wave approximation'** as described in general physics textbooks is entirely adequate and is employed throughout. Appendix A gives a resumé of the notation and basic equations used, together with reminders of the meaning of such terms as path difference and phase difference, and the use of phasor diagrams for summing waves having different amplitudes and phases.

In this chapter we identify, in general terms, some of the physical processes involved in the two main topics. These can be introduced very conveniently in the context of Young's experiment and the phenomena of the Newton's rings type.

The reader will already know that the historical experiment performed by **Thomas Young** in 1801 provided crucial evidence in support of the wave theory of light. We must start by reminding ourselves of some of the details of this experiment and of our present-day interpretation of it.

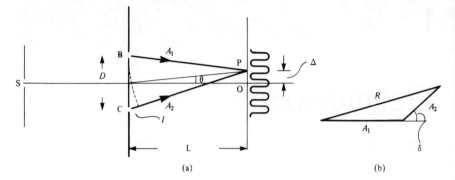

Fig. 1.01 — Young's experiment.

Figure 1.01(a) shows the arrangement. Sunlight passing through a pinhole S illuminated an **aperture mask** (or screen) some distance away, containing two closely separated pinholes B and C. On another screen, about as far away again, fringes of light and dark were observed in the region of geometrical shadow around O. Neither pinhole alone produced the fringes and their existence was interpreted as interference between light diffracted at the two pinholes. It will be recalled that according to **Huygens' principle** as developed by **Fresnel** and **Kirchhoff**, every point of an advancing wavefront is regarded as acting as a source of **'secondary wavelets'** whose envelope forms the profile of the wavefront as it advances. When intercepted by a screen containing an aperture **diffraction** occurs, the wavelets that pass through the aperture having a wavefront envelope that spreads into the region that would be unilluminated shadow according to the **ray theory** of **geometrical optics**. This is indicated in Fig. 1.02(a), which can be thought of as representing one of the apertures in Young's experiment. At any point, P,

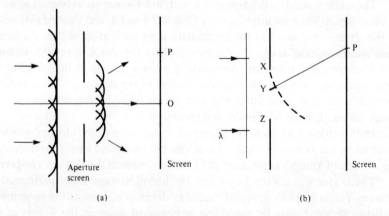

Fig. 1.02 — Diffraction: the spreading of Huygens' wavelets into the geometrical shadow.

say, the illumination is the resultant of the **interference** between the wavelets arriving there from all points across the aperture, arriving with different phases due to the different path lengths they have travelled. The pattern on the screen is of the familiar **Fresnel** type described in the standard texts. The details need not concern us at the moment since, if the pinholes in Young's experiment are small enough, the diffracted light from each, alone, would produce quite a large patch of fairly uniform illumination on the screen. It is the interference between the light diffracted in this way from the two apertures in the experiment that is responsible for the fringes.

Two asides need to be made here. One is to note that the term 'interference' describes what is solely a summation, in accordance with the **principle of superposition,** of displacements where wavetrains overlap; every wavetrain can proceed beyond the region of overlap quite unaffected. In the present context, however, instead of calling the overall pattern on the screen (whatever the number of apertures) an *interference* pattern it is often referred to instead as a *diffraction* pattern in recognition of the physical process by which the light leaves the apertures to arrive at the place where interference occurs. The other point to note is that when the term interference is used in the present context it can be classified as **interference by division of wavefront.** This distinguishes it from **interference by division of amplitude** which occurs in the formation of **Newton's rings,** for example (§1.4).

Before considering the implications of Young's experiment in relation to the subject of this book it will be useful to note some details concerning the fringe pattern itself.

Assume for the moment that monochromatic light, wavelength λ, is used. With the arrangement shown in Fig. 1.01(a), if A_1 and A_2 are the nett separate amplitudes of illumination at P originating from apertures B and C, then the resultant, R, of their superposition at P can be obtained by using the phasor diagram (Appendix A) shown in Fig. 1.01(b), where δ is the phase difference corresponding to the path difference l and is given by

$$\delta = \frac{2\pi}{\lambda} l .$$

The intensity of illumination[†] at P is

$$I = R^2 = A_1^2 + A_2^2 + 2A_1 A_2 \cos \delta$$
$$= A_1^2 + A_2^2 + 2A_1 A_2 \cos \frac{2\pi l}{\lambda} \qquad (1.01)$$

For the sake of generality we will not put $A_1 = A_2$, although that would be justified here.

[†]Flux (power) received per unit area. Alternative terms: **illuminance, irradiance.**

The third term is the 'interference term', and it enables the positions of fringe intensity maxima and minima to be calculated. At points where P is such that $l = 0, \lambda, 2\lambda, \ldots$, the wavefronts diffracted from the two apertures are in phase and we have intensity maxima, with

$$I_{max} = A_1^2 + A_2^2 + 2A_1 A_2 = (A_1 + A_2)^2 \ . \tag{1.02}$$

Between these, where $l = \lambda/2, 3\lambda/2, 5\lambda/2, \ldots$, the wavefronts are in opposite phase (they are 'out of phase'), and there are intensity minima with

$$I_{min} = (A_1 - A_2)^2 \ . \tag{1.03}$$

(For positions of P increasingly away from O the intensity of the pattern is progressively attenuated because the amplitudes of the secondary wavelets diminish with increasing angle. If they did not do so, there would be a wavefront travelling in the backward direction. To deal with this, an **obliquity** (or **inclination**) **factor** — explicit in **Kirchhoff's** analysis — is incorporated in quantitative equations for amplitude/intensity distributions. There would also be a factor allowing for the **inverse square law**, because the distance between P and the apertures varies with the position of P on the screen. In keeping with general practice, such correction terms are omitted from the equations in this book.)

The *spacing* of the fringes, Δ in Fig. 1.01(a), in the central region is given approximately by writing

$$\left. \begin{array}{l} \Delta \approx L\theta_1 \\ D\theta_1 \approx l = \lambda \end{array} \right\} \text{ for first fringe from axis}$$

whence $\quad \Delta \approx \dfrac{L\lambda}{D}$ \hfill (1.04)

It is important to note that the fringe spacing is proportional to λ and inversely proportional to D; also that the actual intensity at any point in the pattern is determined by the amplitude arriving there from each pinhole and the way the two contributions interfere.

There are several respects in which Young's experiment illustrates the types of physical phenomena that are involved in the two main topics of this book mentioned at the beginning of this section. Consider the topic concerned with the spectral and spatial distribution of radiation sources. If the fringes in Young's experiment are to be of good 'visibility' — good clarity — it is essential to use a very small source for the illumination of the apertures. The sets of fringes resulting from light leaving spatially separated points in anything but a very small source would be so displaced relative to each other that the nett fringe pattern would be of low visibility.

Matters concerning fringe visibility permeate much of what lies ahead in this book and we need to note how visibility is defined for analytical purposes, without delay. Figure 1.03 shows the type of effect we have just described.

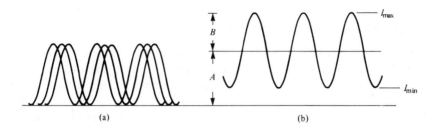

Fig. 1.03 – The smearing of fringes leads to a fringe pattern of the same repeat but reduced visibility.

Displacement of fringes over a small range is depicted in (a), though three distinct fringe patterns are shown whereas in reality there would usually be a continuous spread over whatever is the range (the details would depend on the size and intensity distribution of the source). The nett effect is represented in (b) and the **visibility** is defined as

$$V = \frac{I_{max} - I_{min}}{I_{max} + I_{min}} = \frac{B}{A}. \qquad (1.05)$$

Alternative terms are **modulation** or **contrast**. (Note, however, that 'contrast' is sometimes defined as $(I_{max} - I_{min})/I_{min}$.)

In addition to using a source of small size, the illumination used in Young's experiment needs to be what is loosely called 'monochromatic' if fringes of good visibility are to be observed.

Fringe visibility thus contains information about both the spatial and spectral nature of the source S. It is the extraction of this information from interference effects that will occupy a major part of our thinking and be brought to a head in Chapter 6.

Turning to the other topic, which concerns image formation, radiation is there used as a vehicle for conveying information about an object to the place where the radiation is distributed in such a way as to form an image of the object. With a suitable lens the light leaving the two pinholes in Young's experiment can be used to produce just such an image, of the pinholes. (Similarly, when the pinholes are viewed directly by eye it is the lens in the eye that produces an image on the retina at the back of the eye.) How does the lens function? Why is it that with a large source at S fringes are not seen on the screen yet the insertion of the same lens will still produce an image of them?

The answers to these questions, and the general aspects of the relationships touched on above, between fringe visibility and the nature of the source, involve not only diffraction and interference but also the *'coherence'* of the radiation, to which we must next turn our attention.

1.2 COHERENCE AND LIGHT SOURCES

For interference effects to be observed on the screen in Young's experiment it is necessary that wavetrains of light arriving there from the two apertures (B and C in Fig. 1.01(a)) should overlap and have the same frequency, and that there should be a constant phase difference between them. If these conditions were ideally met, the illumination at the apertures would be said to be **coherent**.

The use of a common light source, S in the figure, for the illumination of both pinholes goes some way to meeting the requirements for coherence. Each pair of wavefronts that leaves B and C has at least originated from a single wavefront from S. And if S were a point source then all wavefronts leaving it would travel a specific distance to B and a specific distance to C, so maintaining a constant phase difference between the diffracted wavefronts leaving B and C.

Conventional light sources are not perfect, however, and they provide illumination that is to a greater or lesser extent coherent — it is **partially coherent**. By its very nature the **photon** (light quantum) emission of light from atoms means that each wavetrain associated with a photon is radiated for a finite time, affecting what is called the **temporal coherence** of the illumination. Furthermore, since any real source is of finite size the wavetrains originate from spatially separated points, affecting what is called the **spatial coherence** of the illumination field given by the source. Both these features of coherence are considered in an introductory way in the following sections, in the context of Young's experiment.

1.2.1 Temporal coherence

The finite duration of the emission of an individual wavetrain of light from an atom means (and we examine it in some detail in §4.6) that it cannot have one single frequency: it would need to be of infinite length for that. Instead, it has a range (albeit narrow) of frequencies, i.e. it has a **'frequency bandwidth'**. Even **laser light** has a finite bandwidth, though this is extremely narrow, with wavetrains some tens of kilometres in length. With conventional, non-laser, sources —commonly referred to as **thermal sources** — thermal vibrations of the radiating atoms, together with other effects, degrade the quality of the light and restrict the time during which a wavetrain can be regarded as approximating to a simple harmonic wave. For these reasons the so-called **monochromatic** light from sources such as gas-discharge tubes is more appropriately called **quasimonochromatic**. **White light** is at the opposite extreme to laser light, with wavetrains so short that no one particular frequency can be identified.

This aspect of the light from any source refers to its temporal coherence, which can be defined qualitatively for the moment as the interval of time during which the phase of the wavemotion changes in a consistently predictable way as it passes a fixed point in space: the longer the time, the greater is the degree of temporal coherence. The length of wavetrain meeting that requirement is the **coherence length**, equal to the product of the coherence time and the velocity of the light. (When the term 'coherence length' is used remember that reference is being made to the spectral purity of the light, not to some aspect of its spatial distribution.)

Even if S in Fig. 1.01(a) were a true 'point source' the effect of finite coherence length means that for points P progressively further away from the axis l may assume a value comparable with the length of the wavetrains. Wavetrains that leave B and C simultaneously (and originate from a single wavetrain from S) would not then fully overlap when they arrive at P, and the visibility of the fringes would consequently be reduced. Further along the screen the fringes could completely disappear to give a continuous level of illumination due to the separate, independent contributions from the two apertures.

As we noted in §1.1, in Young's experiment the finite size of the source also causes reduction in fringe visibility and we consider this in the next section. (In §1.4 an arrangement more specifically sensitive to temporal coherence is described. And the way in whch visibility is related to temporal coherence is considered in §6.4.2.)

1.2.2 Spatial coherence

The radiation processes at separate points in any conventional thermal light source are independent of each other, and in that sense such sources may be thought of as incoherent. But our interest is not so much in the nature of the source itself as in the quality of the illumination field it produces, for example, in a plane at some distance from the source. Thus in Young's experiment we are interested in the extent to which there is a constant phase relationship between B and C (Fig. 1.01(a)) so that interference effects can be observed. We have noted the effect of limited temporal coherence, due to the finite bandwidth of the light from a source. What effect does the finite size of the source have?

Figure 1.04 shows schematically and on a grossly exaggerated scale, the two apertures B and C of Young's experiment, with a light source of width W at distance r. Assume that the light has unlimited coherence length. Light from some point S in the source illuminates the apertures, and interference fringes are produced on the screen. According to the location of S there is a certain delay between the time of arrival of any wavefront at aperture B and its arrival at aperture C. The magnitude of this delay determines the positions of the intensity maxima and minima of the fringes on the screen (not shown in this figure) due to the light from S. If the source could consist of just this single point (as in an idealized Young's experiment) then fringes of maximum visibility would be observed. A real source is of finite size, however, and the fringes

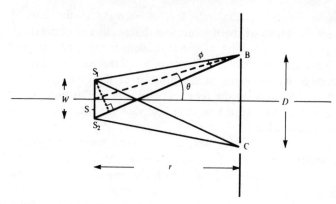

Fig. 1.04 — Spatial coherence. The effect of source size in Young's experiment.

produced by illumination from other points are displaced relative to those due to S. Furthermore, with a conventional thermal light-source, and no matter how great the temporal coherence, the fringe patterns produced by light leaving different places in the source are purely additive in terms of intensities (i.e. there is no interference between them) since they are quite unrelated. Light from the whole source therefore produces a spread of fringes as illustrated in Fig. 1.03, with a consequent reduction in the visibility of the fringe pattern.

In general, then, the illumination field from a finite source, even if truly monochromatic, is again partially coherent, this time spatially — owing to the spatial distribution of the source. And the degree of coherence between two points in the field is physically manifest in the visibility of interference fringes produced by light from those two points. The actual relationship between spatial coherence and the fringe visibility is dealt with in §6.4.

For practical purposes, one can say that a high degree of spatial coherence is obtained at the aperture screen in Fig. 1.04 if BS–CS only changes by a small fraction of a wavelength for all positions of S from S_1 to S_2. This can be expressed as

$$(BS_2 - BS_1) - (CS_2 - CS_1) \ll \lambda .$$

As W must obviously be very small compared with r, the above approximates to

$$2W\theta \ll \lambda .$$

Since $r\theta \approx D/2$ and $r\phi = W$ we then have

$$\phi \ll \frac{\lambda}{D} , \qquad (1.06)$$

which tells us that for the illumination at the pinholes B and C to be spatially coherent, the angle ϕ (in radians) subtended by the source at the pinholes must be appreciably less than λ/D. Conversely, we can say that in the plane of the pinholes the illumination is effectively coherent spatially over a distance D equal to a small fraction of λ/ϕ, the fraction being specified according to any particular experimental requirement. That particular value of D is called the **(transverse) coherence width** of the illumination in the plane concerned. Given sufficient information about the source the **coherence area** in a plane at a specified distance from the source can be defined in the same way.

It is interesting to pause here and realize that no matter how drastically the coherence of an illumination field may differ from one point to another, such differences are not visible to the eye; the eye is only sensitive to intensity.

To return to the reduction in fringe visibility that occurs as source size increases, this provides the basis of *Michelson's* **stellar interferometer** method for measuring the angular diameters of stars (the angles they subtend at the Earth) that are too small to be measurable with a telescope in the conventional way. This is outlined in Chapter 6 where we also see how the actual way in which the visibility changes as the separation between the two apertures is altered reveals information about the brightness distribution of the source.

Finally, we should note that laser light sources are an exception with respect to spatial coherence just as they are with temporal coherence (§1.2.1). There is spatial coherence right across a laser beam (and its width can be increased for practical purposes by means of a suitably designed lens system — a **beam expander** — without loss of coherence). Closely similar to laser light, from the point of view of spatial and temporal coherence, are the **radio waves** emitted by radio transmitters.

1.3 OPTICAL IMAGE FORMATION

We are now in a position to identify and appreciate some of the general aspects of optical image formation, before embarking on their detailed study leading to Chapter 5.

In §1.1 we noted that in Young's experiment the introduction of a lens would enable an image of the two pinholes to be formed. It is evident, then, that diffraction at the pinholes is the first step in the formation of their optical image, the second step being the 'recombination' of that light by the lens to form the image.

We also recalled in §1.1 that fringes are not observed in Young's experiment if a large light source is used. Yet it is a common experience that a lens would still produce an image of the apertures.

Does the lens act differently in these two cases? The short answer is no, for reasons that can fairly readily be appreciated in qualitative terms.

Consider the first case, when Young's experiment is performed in the classical way with what we can now describe as reasonably coherent illumination at the aperture screen. The source is small enough to ensure that the same phase relationships exist between every pair of diffracted wavefronts leaving the apertures B and C. Every pair contributes to the same set of fringes, and the intensity maxima and minima in the fringe pattern can be located by reference to Eqn (1.01).

In the second case, using a large source, there is no spatial coherence in the illumination at the aperture screen (we can assume that the temporal coherence is adequate in both cases). As we have seen, fringes resulting from wavefronts leaving one point in the source are displaced with respect to fringes produced by wavefronts originating from other points in the source. If the source is large enough the nett result is a relatively uniform patch of illumination on the screen. Even so, fringes are indeed formed as in the first case, but they are instantaneous and their positions continually change, giving uniform illumination when viewed over even the shortest practicable time. That the information is there, with individual phase relationships still existing in the light that will enter the lens when this is inserted, is evident from the fact that it is still possible to form an image of the aperture with the lens. (When coherent illumination is used the stability of the phase relationships makes it possible under some circumstances to record the full information, including phases, in the wavefronts leaving the aperture screen; this is the basis of holography (§5.4).)

The above comments concerning the imaging under coherent illumination ('coherent imaging' for short), of an object in the form of the pinhole mask in Young's experiment, apply equally to (i) more complicated object masks such as the 35-mm transparencies used in slide projectors, (ii) opaque objects illuminated with conventional thermal light, and (iii) 'self-luminous' objects which are luminescent (e.g. a TV picture) or incandescent (e.g. infra-red photography of hot objects). The same *instantaneous* phase relationships exist in each of these categories, just as we have described above.

In this wider context of imaging the term **'scattering'** is often used instead of **'diffraction'**. There is some confusion in making a distinction between the two in the literature but the details do not affect our deliberations and we can regard the terms as synonymous.

There are two aspects of image formation that now need to be noted. One is concerned with the formation of a Fraunhofer-type diffraction pattern in the back focal plane of the imaging lens, the other with the effect that the finite aperture of the lens has on the image. (The matter of lens aberrations can wait till Chapter 5.)

1.3.1 Fraunhofer diffraction

Consider Fig. 1.05 which depicts an object mask containing two very small pinhole apertures, B and C, uniformly illuminated with quasimonochromatic light from a distant source. Plane waves arrive at normal incidence to the mask, and spherical wavefronts spread from B and C. This is the same as in Young's experiment except that in addition we now have a lens producing an image of

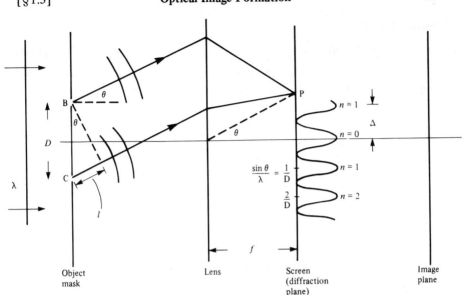

Fig. 1.05 — Diffraction as an intermediate step in image formation.

the pinholes in a plane located as shown in the figure. It is the back focal plane of the lens, however, that is of immediate interest. Consider any point P, at θ to the lens axis; only contributions travelling in direction θ from B and C are brought together there, to interfere (cf. Young's experiment where the interference at P in Fig. 1.01 is between light travelling in *different* directions from the apertures). We shall see that this particular diffraction pattern (defined below as of the Fraunhofer type), formed in the back focal plane of the imaging lens, is a particulary important intermediate step in the formation of the image produced by the lens. It provides valuable insight into the final stage of image formation, and an opportunity of special and unique significance for modifying the image. These are among the topics of Chapter 5, but here we will examine some of the features of the pattern formed in the example we have just described. First, however, note that for such Fraunhofer diffraction patterns to be experimentally accessible it is necessary to arrange that the static phase relationships provided by coherent illumination exist — see the comments in the previous section about the difference between coherent and incoherent imaging. Until Chapter 5, where those differences are again discussed, we shall — unless otherwise stated — make the assumption that coherent conditions obtain.

To calculate the resultant of the two contributions at P in Fig. 1.05, it is necessary to take into account the phase difference, δ, between them — as we did with Young's experiment — due to the path difference l. We have

$$\delta = \frac{2\pi l}{\lambda} \qquad (1.07)$$

where, this time,

$$l = D \sin \theta \ . \tag{1.08}$$

The intensity, I, at P can be calculated with a phasor diagram in a similar way to that in Fig. 1.01. This gives (cf. Eqn (1.01))

$$I = A_1^2 + A_2^2 + 2A_1 A_2 \cos \frac{2\pi l}{\lambda}$$

where l is now given by Eqn (1.08).

The two amplitudes are equal since the same direction, θ, is involved for both contributions. Putting $A_1 = A_2 = A$ the above equation becomes

$$\begin{aligned} I &= 2A^2 \left[1 + \cos\left(\frac{2\pi}{\lambda} D \sin \theta\right)\right] \\ &= 4A^2 \cos^2\left(\frac{\pi}{\lambda} D \sin \theta\right) \end{aligned} \tag{1.09}$$

The pattern is commonly referred to as consisting of '$\cos^2$ **fringes**'.

In the direction of a fringe intensity maximum the illumination from each aperture is in step and

$$l (= D \sin \theta) = n \lambda \tag{1.10}$$

where n, the order of diffraction (or interference), is an integer or zero. The intensity is then

$$I_{\max} = 4A^2 \ .$$

It should again be noted that although the directions of the maxima are determined by D and λ, their magnitudes are controlled by the value of the amplitude transmitted in those directions by the individual apertures (dealt with in §2.3).

Between the maxima there are minima, which are zero only if (as is usual) $A_1 = A_2$.

The fringe spacing, Δ, is readily calculated with the aid of Fig. 1.05. If the first-order maxima are at $\theta = \pm \theta_1$ say, and if θ_1 is small (the figure is of course grossly exaggerated) we have

$$\Delta \approx f \theta_1$$

$$D \theta_1 \approx \lambda$$

whence $\quad\quad\quad \Delta \approx \dfrac{f\lambda}{D} \ . \tag{1.11}$

Regardless of the shape, number, etc., of apertures in the mask, the patterns formed in this way are all described as being of the **Fraunhofer** type. The arrangement is shown again in Fig. 1.06, with an object mask consisting of a single aperture of finite size; the details of the Fraunhofer pattern this gives are dealt with in §2.2. In all examples of Fraunhofer diffraction there is a linear variation of the optical path length travelled by diffracted light from points across the object screen to any particular point in the pattern. Thus in Fig. 1.06 the optical path difference YP−XP=YW is proportional to XY. In contrast, the corresponding variation in Fig. 1.02 is non-linear, and patterns formed under those conditions are of the **Fresnel** type.

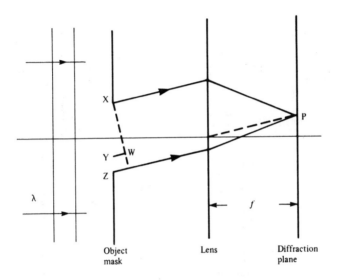

Fig. 1.06 − Fraunhofer diffraction.

Since with a Fraunhofer diffraction pattern the same pattern would be obtained at infinity if the lens were removed, an alternative description often used is **far-field diffraction**. In contrast, Fresnel diffraction is then referred to as **near-field diffraction**, though it must be stressed that a wide variety of pattern falls into the Fresnel (near-field) category, with the Fraunhofer type at one extreme. For example, when Young's experiment is performed with both the source and the screen (on which the fringes are observed) at appreciable distances from the aperture mask the pattern is practically indistinguishable from the Fraunhofer type. If the distances are substantially reduced (as suggested in the exaggerated scale of Fig. 1.01(a)) the Fresnel pattern would be quite different.

It is not necessary here to go into the vast subject of the methods of illumination used in optical microscopy. But it is useful to note in Fig. 1.07 some of the arrangements used specifically for observing and recording Fraunhofer diffraction patterns − often for 'optical filtering' purposes (Chatper 5). Of

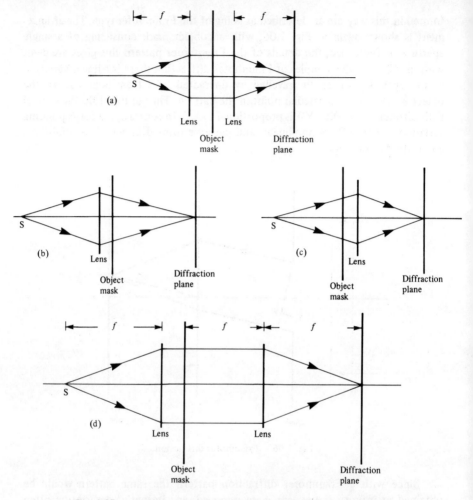

Fig. 1.07 — Arrangements for forming Fraunhofer diffraction patterns.

these, (a) is another way of providing the same illumination as in Figs. 1.05 and 1.06, whilst (b) and (c) produce almost the same patterns as (a). When accurate details are required concerning both phase and amplitude, the requirements are more stringent. This is due to the imaging properties of lenses, which when fully taken into account show that the arrangement shown in (d) is required. Here the object screen is located in the front focal plane of the lens in whose back focal plane the diffraction pattern is formed.

In the remainder of this book it can be assumed, unless otherwise stated, that the various diffraction patterns discussed are of the Fraunhofer type or that they approximate sufficiently to it for the differences to be negligible for the purpose concerned — the so-called far-field (or plane-wave) approximation.

As already mentioned, the role of diffraction in an optical imaging system as an intermediate step in image formation is an important topic of Chapter 5. To that end it should be noted that the Fraunhofer diffraction pattern given by an object (such as the aperture masks considered here) is observed in the plane where the image of the source as object is formed (cf. Fig. 1.07); the popular example is the observation of the pattern seen when a street lamp is viewed through a net curtain. Using the term **'conjugate'** in the same sense as in geometrical optics (where one refers to an image as being formed in the plane conjugate to the object), the Fraunhofer diffraction pattern is said to be formed in the plane conjugate to the light source. (This is summarized diagrammatically in Fig. 5.05.)

1.3.2 Lens aperture

Since the wavefront system leaving an object is the only information on which the formation of its image is based it is to be expected, and is indeed found, that the more of the wavefront system that is allowed to enter the imaging lens the better is the quality of the image. Rephrased, this is the well-known axiom that the larger the aperture of the lens the better is the definition of the image (assuming aberrations are not a limiting factor).

An alternative, equivalent, model considers how the finite aperture of the imaging lens would degrade the imaging of each point in the object individually. The reader will recognize that this uses the historical work on the resolving power of telescopes, where it will be recalled that the image of a star (a close approximation to a point source) is blurred by diffraction at the lens aperture into a disc surrounded by rings — the **'Airy pattern'** named after Sir George Airy the Astronomer Royal who in 1835 worked out the details of the pattern (§2.3). The dimensions of the Airy pattern are inversely proportional to the diameter of the diffracting aperture, so that only if the aperture were infinitely large would each object point be imaged as a point.

Both these approaches to imaging have their uses, and they are dealt with in some detail in Chapter 5.

1.4 INTERFERENCE BY DIVISION OF AMPLITUDE

Newton's rings and the fringes seen in thin films such as soap bubbles, oil on water, etc. are the result of interference occurring when light is partially reflected at two (or more) successive boundaries between media of different refractive indices. If a wavetrain of incident light is partially reflected at the first interface (air/oil in the case of oil on water) then a reduced amplitude of the same wavetrain is transmitted and then partially reflected at the next interface (oil/ water). Interference occurs if the two reflections are brought together, as when viewed by the eye for example, with a result that depends on the path difference that has been developed between them owing to the separation of the two reflecting surfaces. (In white light, colour effects are observed when the path difference — a function of film thickness and angle of viewing — is such that there is constructive interference for some wavelengths and destructive interference for others.)

In contrast to interference involved in diffraction, where wavefronts are 'divided' by apertures, the above effects are classified as interference by division of amplitude and devices based on this type of interference are 'amplitude-splitting interferometers'. The example depicted in Fig. 1.08(a) is of interference between partial reflections at the two surfaces of a thin parallel plate. Each

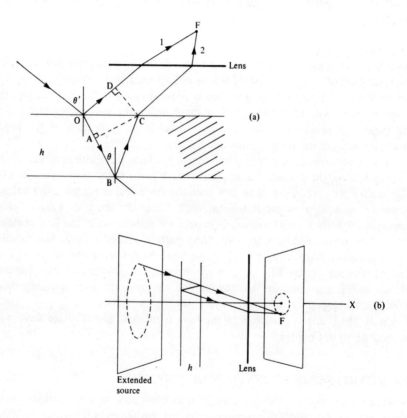

Fig. 1.08 – Interference by division of amplitude.

incident wavetrain is partially reflected at the two interfaces, air/glass at O and glass/air at B. If μ is the refractive index of the glass then the optical path difference, l, between the two reflections (1, 2) at O and C is given by

$$l = \mu(\text{OB} + \text{BC}) - \text{OD}$$

Since $\mu OA = OD$ this reduces to

$$l = \mu (AB + BC)$$
$$= \mu BC (\cos 2\theta + 1)$$
$$= \frac{\mu h}{\cos \theta} 2\cos^2 \theta$$
$$l = 2\mu h \cos \theta . \tag{1.12}$$

Constructive interference between reflections 1 and 2 occurs if

$$2\mu h \cos \theta = (m + \tfrac{1}{2}) \lambda \tag{1.13}$$

where m is an integer or zero. The $\tfrac{1}{2}\lambda$ on the right side of this equation is to allow for the phase change of π involved in the reflection at O (cf. Stokes's treatment of reflection and refraction (Appendix B)).

If the incident light is not restricted in Fig. 1.08(a) to the plane of the figure concentric circular interference fringes can be observed, as shown schematically in (b). This is because the above condition for an intensity maximum, such as at F, is equally fulfilled by all points on a circle through F with its centre on the axis X.

In view of the comments in §1.2 about coherence it is evident that since h can be appreciable, the effect of temporal (rather than spatial) coherence (and thereby the spectral composition of the illumination) on fringe visibility with this arrangement can be considerable. To study this in practice a parallel air-gap of adjustable width is used. In Chapter 6 we meet such an arrangement in the guise of Michelson's spectral interferometer and we see how the variation of fringe visibility with plate separation is related to the spectral composition of the light by a Fourier transform. As mentioned earlier, it forms the basis of some modern methods of spectroscopy.

2

Fraunhofer diffraction

2.1 INTRODUCTION

In this chapter details of Fraunhofer diffraction patterns given by some simple apertures and gratings are described in preparation for subsequent chapters. Much of this should already be familiar to the reader, but certain features of particular relevance later need to be identified and highlighted.

The chapter concludes with a look at the close analogy between optical diffraction and the diffraction of X-rays by crystals. The indirect way in which the 'imaging' of atoms in crystals can be achieved by using X-ray diffraction data, indirect because lenses for X-rays are impracticable, will then be briefly described in Chapter 5 as an application of the theory of image formation.

2.2 SINGLE-SLIT PATTERN

Figure 2.01(a) depicts a section of a slit of width a, and of length l perpendicular to the plane of the figure. The slit is uniformly illuminated with monochromatic light, wavelength λ, fulfilling the requirements of §1.2 for coherence, and the arrangement is such that plane wavefronts arrive at the slit at normal incidence. The Fraunhofer diffraction pattern given by the slit is then formed in the back focal plane of the lens. We assume that $l \gg a$ so that events in planes parallel to that of the figure can be assumed to be the same: such a pattern is a **'1-dimensional' diffraction pattern**. Its details can be derived using the Huygens' wavelets model.

Imagine the slit to be divided, parallel to its length, into a large number of narrow strips, all of the same width. The whole slit is uniformly illuminated, the strips are of equal area, so the amplitude of the contribution from each strip in a given direction θ is the same. Calculation of the combination of all the contributions from across the entire slit is most easily done with the help of a phasor diagram. The phase of the light from the central strip at C can serve as a convenient reference baseline in the phasor diagram. In Fig. 2.02(a) the phasor for the light from this strip is labelled C'. The phase of the light from the adjacent

[§ 2.2] **Single-Slit Pattern** 29

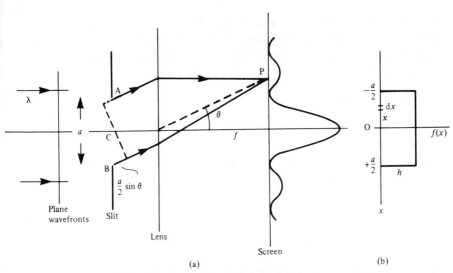

Fig. 2.01 – Single-slit Fraunhofer diffraction.

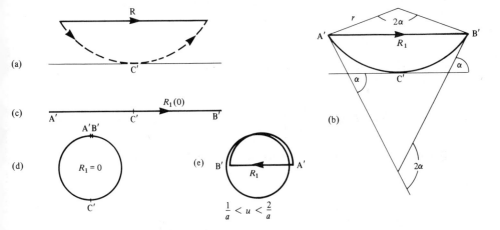

Fig. 2.02 – Single-slit diffraction: phasor diagrams.
(The slight displacement in (e) is to show that the overall phase difference is 3π.)

strip below C in Fig. 2.01(a) lags behind that from C, and that from the adjacent strip above C is ahead of that from C. In the phasor diagram these are added to C' accordingly, as are pairs from strips further and further away from C, until the contributions from all the strips have been included. R is the resultant given by this construction, but to obtain an accurate value we need a phasor diagram which represents the limit when the strips become infinitesimally narrow. The transition is easily made. First we note that the present phasor diagram resembles part of a regular polygon because (i) the phasors are of equal length, and (ii) they are rotated through equal angles (owing to the linear variation of phase between the contributions from successive strips). The light from A in Fig. 2.01(a) has a phase ahead of that from C by an amount

$$\alpha = \frac{2\pi}{\lambda} \cdot \frac{a}{2} \sin\theta \qquad (2.01)$$

and that from B lags behind that from C by α. For light leaving a position halfway between C and B the phase lag is $\alpha/2$, and so on. In the limit when the strip widths are reduced to zero the phasors form an arc of a circle of radius r, as shown in Fig. 2.02(b), and the resultant amplitude, R_1, is given by

$$R_1 = 2r \sin\alpha . \qquad (2.02)$$

The subscript is to remind us later that this resultant refers to a single slit.

The arc $A'C'B'$, of length $r.2\alpha$, is equal to the total length of all the phasor elements stretched out in a straight line, as in (c). Now this straight line represents the phasor diagram for the $\theta = 0$ direction, since in that special direction (cf. Fig. 2.01(a)) there are no path differences and all the contributions are in phase with that from C. We may therefore write

$$R_1(0) = r . 2\alpha$$

where $R_1(0)$ denotes the amplitude in direction $\theta = 0$.
Substitution into Eqn (2.02) gives

$$R_1 = R_1(0) \left[\frac{\sin\alpha}{\alpha} \right] . \qquad (2.03)$$

The features of the term in brackets are shown in Fig. 2.03(a).

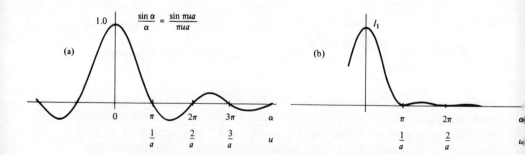

Fig. 2.03 — Single-slit diffraction pattern.
(a) amplitude (sinc function), (b) intensity.

[§2.2] Single-Slit Pattern

The experimentally observed intensity pattern is given by

$$I_1 = R_1^2$$

as shown in Fig. 2.03(b) and Fig. 2.12(a).

The variable α is a function of $(\sin \theta)/\lambda$ and a (cf. Eqn (2.01)). Interpretation of the diffraction pattern is simplified if $(\sin \theta)/\lambda$ is replaced by the single variable

$$u = \frac{\sin \theta}{\lambda} \quad . \tag{2.04}$$

Eqn (2.03) then becomes

$$R_1 = R_1(0) \left[\frac{\sin \pi u a}{\pi u a} \right] \quad . \tag{2.05}$$

The bracketed term in this equation is an example of a **sinc function**, of general form $(\sin \pi x)/\pi x \equiv$ 'sinc x', which occurs widely in physics.

From Fig. 2.03 there is seen to be a reciprocal relationship between the scale of the pattern in terms of u and the slit width a. Now u has the dimensions of reciprocal length, and we are therefore led to the concept of describing Fraunhofer diffraction patterns as existing in 'reciprocal space'. This will be developed further when we consider multiple-aperture patterns (§2.4 and §2.5) and it will take on additional meaning when we encounter the Fourier aspects of diffraction (Chapters 3 and 4).

Meanwhile it is necessary to note in Fig. 2.03(a) the variation with u, of the phase of the amplitude pattern. As u starts to depart from zero the resultant amplitude decreases, corresponding to the curving of the phasor diagram (b) in Fig. 2.02. But the resultant remains parallel to the phase baseline, i.e. its phase is that of the illumination in direction $u = 0$. When $u = 1/a$ the phasor diagram has formed a complete circle (d) and the resultant amplitude is zero. [$u = 1/a$ means that $a \sin \theta = \lambda$, i.e. one whole λ path difference between the contributions from A and B (Fig. 2.01(a)). This seems to conflict with a zero nett amplitude from the whole slit, until one notes that it means that there is therefore a $\tfrac{1}{2}\lambda$ path difference causing cancellation between the contributions from A and C, and similarly between those from just below A and just below C, and so on right across the slit.] After that, as u increases to $u = 2/a$ the phasor diagram (Fig. 2.02(e)) exceeds one full circle, and there is again a resultant parallel to the baseline. However, the phase is now π, as manifest in the negative value of the amplitude in Fig. 2.03(a). Further zeros and phase reversals follow at $u = 2/a, 3/a, 4/a, \ldots$, the phasor diagram forming ever-decreasing circles of more and more turns.

The amplitude distribution of illumination across any aperture system such as the present slit — the **amplitude transmittance** of the system — is represented by the **'aperture function'**. In Fig. 2.01(b) the aperture function takes the form of a simple **rectangle** (or **'top hat') function**, $f(x)$, which has a constant value, h, over the whole width of a single slit, and is zero elsewhere. (In general, $f(x)dx$ is the amplitude of wave-motion originating from element dx at x.)

In the context of this book the 'aperture function' of any system of apertures (not necessarily just one aperture as here) can usefully be thought of as representing what one can call the **(optical) 'structure'** of that system.

2.3 CIRCULAR APERTURE

The Fraunhofer diffraction pattern formed by a circular aperture is of considerable importance in connection with the performance of most optical instruments. Unfortunately, the details of the pattern are difficult to derive by the phasor method used in the previous section for a slit aperture. The reason is that the strips into which the aperture is imagined to be divided are not now all the same length (cf. aperture functions in Fig. 2.05(a)). They progressively increase and then decrease in size across the aperture, and the phasor diagram would not form part of a regular polygon. This example is more easily solved analytically and details are to be found in some of the standard textbooks. The pattern (Fig. 2.04(a)) consists of a central disc surrounded by concentric circular fringes and is known as the **Airy pattern**, after Sir George Airy, the Astronomer Royal, who derived the details in 1835.

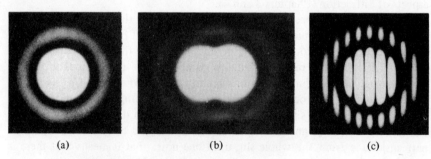

Fig. 2.04 — Circular-aperture diffraction patterns.
(a) Single aperture (the Airy pattern).
(b) Two apertures illuminated separately, incoherently.
(c) Two apertures, illuminated by a common, coherent source.
((a)(c) reproduced by permission, from Harburn *et al.* (1975), *Atlas of optical transforms*. Bell, London.)

The amplitude variation along a diameter of the Airy pattern is shown by the full line in Fig. 2.05(b). Though closely resembling the pattern for a slit aperture (shown by the broken line) it is described by a Bessel function instead of the sinc function. For us, an important difference between the two is that the

Circular Aperture

first intensity zero from the central peak of the Airy pattern does not occur until

$$a \sin \theta \approx 1.22\lambda \qquad (2.06)$$

where a is the diameter of the aperture. Subsequent zeros are also displaced.

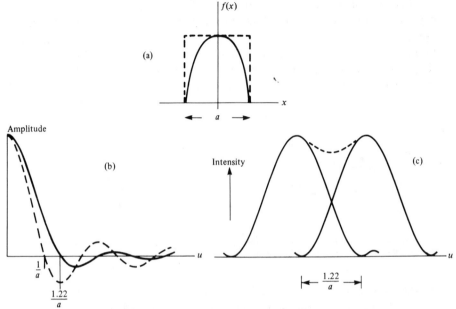

Fig. 2.05 — (a) Aperture functions, $f(x)$: solid line, uniformly illuminated circular aperture of diameter a; dashed line, uniformly illuminated slit of width a.
(b) Amplitude diffraction patterns: solid line, circular aperture of diametre a; dashed line, slit of width a.
(c) Rayleigh criterion (circular aperture patterns).

An immediately obvious example of the formation of this pattern occurs when an astronomical telescope is used to image a star, and it was in this context that Airy derived the details of the pattern. A star is effectively at an infinite distance from an observer on Earth and its image is therefore formed in the back focal plane of the telescope objective, where it is examined with the aid of the eyepiece. Since its angular diameter (the angle subtended at Earth by its diameter) is extremely small the image should be almost a point. However, the wavefronts from the star are interrupted by the limited aperture of the objective, diffraction occurs and it is the Fraunhofer diffraction pattern of this aperture that is seen as the image of the star. As an example of Fraunhofer diffraction this corresponds to the arrangement shown in Fig. 1.06, with the aperture of the objective itself acting as the diffracting mask.

No matter how free from imperfections an astronomical telescope objective (lens or mirror) is, what is observed at best is not a point image of the star, but the Airy intensity pattern produced by the aperture of the telescope objective (such a lens is **'diffraction limited'**). In the wider context of Chapter 5 this pattern — the response of the system to a point ('impulse') input — is the **point-spread function (PSF)** of the system.

Eqn (2.06) shows how the diameter of the central disc (the **Airy disc**) depends on the diameter of the aperture and the wavelength of the light. It is the size of this disc that determines the limiting resolution of a telescope. Consider the image of two stars of small angular separation θ (Fig. 2.06). As they are mutually incoherent sources the image consists of two independent Airy intensity patterns. Resolving the presence of two stars depends, therefore,

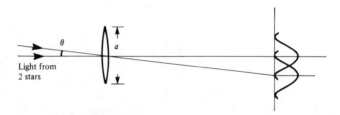

Fig. 2.06 — The astronomical telescope. The Rayleigh criterion applied to the imaging of two stars of small angular-separation.

on the sizes of the Airy discs and the extent to which they overlap. The conventional specification for the limiting condition, the **Rayleigh criterion**, is the separation shown in Figs. 2.04(b) and 2.05(c). According to this the two patterns are resolved if the centre of the Airy disc of one pattern is superimposed on the first dark ring from the centre of the other. This gives a dip in the total overlap intensity of about 20% between the two peaks (assumed equal in magnitude for the present purpose). A dip of this magnitude though rather arbitrary, is a convenient criterion of resolution for many purposes.

On the basis of the Rayleigh criterion, Eqn (2.06) then gives the angular resolution limit, θ for a **telescope** of aperture a as

$$\theta \approx 1.22 \lambda/a \qquad (2.07)$$

or $\quad \theta \approx \dfrac{0.15}{a_{\text{metres}}} \quad$ secs or arc

as a rough guide for yellow light.

(To try to improve the ability of a telescope to resolve very close images of stars, the circular fringes surrounding an Airy disc can be suppressed. This is done by 'apodization' — modifying the aperture function of the objective by using, for example, a glass plate whose transmission properties vary from point to point in a prescribed way. The price paid for restricting the transmission of

[§2.3]   **Circular Aperture**   35

light through the objective is, however, liable to be the broadening of the central disc itself.)

The same factors determine the resolution limit of a terrestrial telescope or camera. Under normal conditions of illumination every point of a terrestrial object scatters light independently of its neighbours and is therefore independently imaged. This is arguably the same as if a cluster of stars were being imaged. For this reason the term 'self-luminous object' tends to be used with some licence in both contexts as a short way of referring to objects imaged under incoherent conditions. With a terrestrial telescope or camera, the image corresponding to every object point as source again consists not of a point but of the diffraction pattern of the aperture of the objective (cf. §1.3.1). (We shall not consider the role of the **eyepiece** in image formation with a telescope — or compound microscope (below) — since it is a relatively low-powered second-stage component and not a prime source of imperfections.)

A similar situation applies when an optical **microscope** (simple or compound) is used to examine a 'self-luminous' object under conditions like those considered above. Using the Rayleigh criterion as for the telescope, the minimum separation s in Fig. 2.07(a), of two points in an object that can be resolved by a microscope, is given by

$$s \approx 0.6\lambda/\sin i \qquad (2.08)$$

where i is the semi-angle subtended by the objective at the object. If the object space is filled with a medium of refractive index μ — using an 'oil-immersion objective' — then the wavelength of light there is reduced to λ/μ and the above expression becomes

$$s \approx 0.6\lambda/\mu\sin i \qquad (2.09)$$

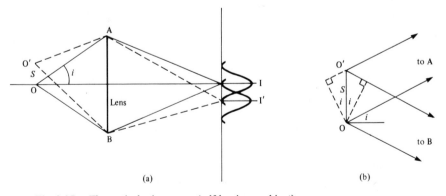

Fig. 2.07 – The optical microscope (self-luminous object).
(a) The Airy-pattern images of two object points O and O', distance s apart, are resolved according to the Rayleigh criterion, by the objective lens, if $O'B - O'A = 1.22\lambda$, since $OA = OB$. From (b): $O'B = OB + s \sin i$, $O'A = OA - s \sin i$, whence $s = 0.6\lambda/\sin i$.

where $\mu \sin i$ is the **numerical aperture** of the microscope objective. The maximum possible value of i is 90°, giving the **'microscopic limit'** as approximately $\tfrac{1}{2}\lambda/\mu$.

With optical imaging systems in general there is usually a limiting aperture that controls the light-gathering power of the system. This **'aperture stop'**, often located between the various lens components of the systems, inevitably gives rise to diffraction as described above. As seen from the object (i.e. source) this aperture is called the **entrance pupil**, and as seen from the image it is called the **exit pupil**. In the language of instrumental optics the pupils are thus the images of the aperture stop as formed in the object and image spaces. And the aperture function, already defined in §2.2, when expressed in terms of the image-space coordinate system, is called the **(exit) pupil function**.

2.4 DOUBLE APERTURE

2.4.1 Two slits

The derivation of the details of the Fraunhofer diffraction pattern obtained with a double-slit arrangement provides a useful introduction to the multiple-slit pattern (§2.5), where exactly the same principles are involved.

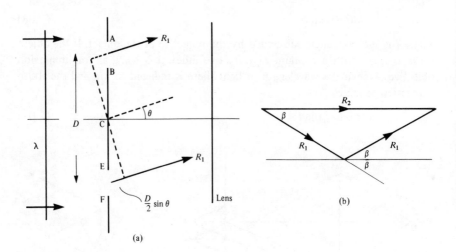

Fig. 2.08 — Double-slit diffraction.

The arrangement shown in Fig. 2.08(a) is similar to that for the single slit, but with two slits parallel to each other and with their centre-lines distance D apart. The spatial pattern of complex amplitude leaving each slit is exactly the same as before but the nett result in any direction θ depends on the path difference between the two contributions in that direction. For example, only when θ is such that the diffracted light from the two slits is in step is there complete

[§2.4] Double Aperture

reinforcement. From the figure it follows that this condition is fulfilled when

$$D \sin \theta = n \lambda \tag{2.10}$$

i.e. $\dfrac{\sin \theta}{\lambda} = u = \dfrac{n}{D}$ (using 2.04)

where n, the diffraction order, is zero or an integer. In all other directions there is destructive interference to different extents.

A phasor diagram leads to an expression for the whole pattern. Relative to a phase reference origin chosen for convenience at C in the figure, slit AB diffracts in direction θ with a phase $+\beta$ and slit EF diffracts with a phase $-\beta$, where

$$\beta = \frac{2\pi}{\lambda} \frac{D}{2} \sin \theta \tag{2.11}$$

i.e. $\beta = \pi u D$

From the phasor diagram (Fig. 2.08(b)) the resultant amplitude is

$$R_2 = 2R_1 \cos \beta .$$

With R_1 given by Eqn (2.05) and β from Eqn (2.11), this gives

$$R_2 = 2R_1(0) \left(\frac{\sin \pi u a}{\pi u a} \right) \cos \pi u D . \tag{2.12}$$

Omitting the constant term this expression is constructed in Fig. 2.09. The alternating phases in the amplitude pattern are understood if phasor diagrams like the one used above are drawn for various values of β. The final graph in Fig. 2.09 shows the intensity distribution, obtained by squaring the amplitude pattern, viz.

$$I_2 = 4I_1(0) \left(\frac{\sin \pi u a}{\pi u a} \right)^2 \cos^2 \pi u D . \tag{2.13}$$

An example of the observed pattern is shown in Fig. 2.12(b).

There are several important points to note in connection with the double-slit pattern. As Eqn (2.13) and Fig. 2.09 show, the pattern is the product of two diffraction 'terms': the single-slit pattern is multiplied by a $\cos^2$ fringes pattern associated with interference between the diffracted light from the two slits.

The maxima of the $\cos^2$ fringes correspond to directions in which there is constructive interference between light from the two apertures. We then have $D \sin \theta = n\lambda$, i.e. $u = n/D$ (Eqn (2.10)), the $\cos^2$ term is unity, and the observed intensity is — as one expects — the square of the sum of the separate amplitudes from the two slits. In a similar way, between these directions there is destructive

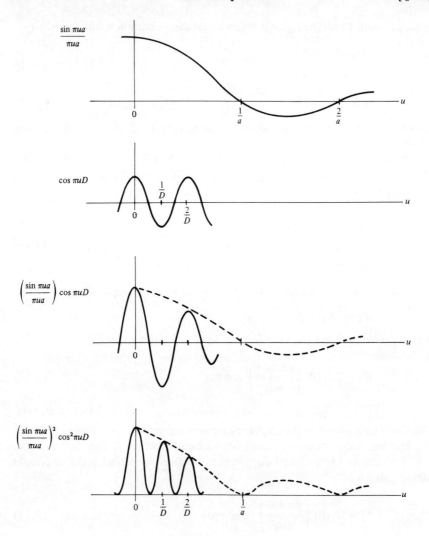

Fig. 2.09 – Double-slit diffraction pattern.

interference, the $\cos^2$ term is zero and the observed illumination is zero. Thus the observed maxima and minima are in directions determined by the slit separation D. Note that their intensities, however, are determined by the amplitudes of the light diffracted from the slits in those directions. In this sense the observed diffraction pattern can be regarded as an **'amplified sampling'** of the single-slit pattern, the sampling being restricted to directions determined by the slit separation. This is evident in Figs. 2.09 and 2.12(b) where the single-slit term is seen to be the envelope of the $\cos^2$ fringes. Inserting the condition $u = n/D$ for reinforcement into Eqn (2.13) gives

$$(I_2)_{max} = 4I_1(0) \left[\frac{\sin \frac{\pi n a}{D}}{\frac{\pi n a}{D}} \right]^2 \tag{2.14}$$

and the term in brackets is what can be described as a *sampled sinc function*.

In accordance with the interpretation of u as a parameter in reciprocal space (§2.2), this sampling of the single-slit pattern can be regarded as taking place in that space. We can say that the double-slit pattern is a sampling in reciprocal space of the single-slit pattern; and the sampling is at values of u determined by, and reciprocally related to, the slit separation D, i.e. at $u = n/D$. Thus the smaller the value of D the wider is the separation of the fringes.

2.4.2 Two circular apertures
The derivation of the diffraction pattern produced by two circular apertures distance D apart is closely analogous to that from two slits. The full details need not be worked out here. As illustrated in Fig. 2.04(c) we again have the single-aperture diffraction pattern (here the Airy pattern) multiplied by the same $\cos^2$ term as above – i.e. the term that is associated with the separation (D) of two apertures. Note in Fig. 2.04 the difference between (b) and (c). It will be recalled that the pattern in (b) consists of two quite separate Airy *intensity* patterns, produced incoherently by different sources. In (c), where there is coherence, it is the phasor summation of the two Airy *amplitude* patterns that is involved.

2.5 N-SLIT GRATING
The same phasor method (Fig. 2.10) as that used for the double-slit in §2.4.1 readily gives the details of the diffraction pattern obtained with the conventional type of multiple slit 1-dimensional plane transmission grating used in the laboratory.

If the grating consists of N slits there are N phasors each of length R_1, this being the amplitude from each individual slit as before (see Eqn (2.05)). Again, there is an angle of 2β between successive phasors, where $\beta = \pi u D$ (Eqn (2.11)). From Fig. 2.10 we see that the ends of the phasors lie on a circle of radius $r' = R_1/2\sin \beta$. The resultant, R_N, for N slits is therefore

$$R_N = 2r'\sin N\beta = R_1 \frac{\sin N\beta}{\sin \beta} \quad .$$

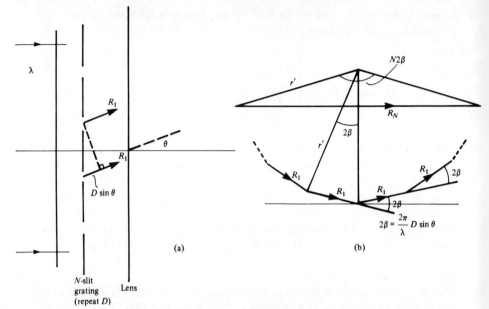

Fig. 2.10 – N-slit diffraction.

Substituting for β gives

$$R_N = R_1 \frac{\sin N\pi uD}{\sin \pi uD}$$

where D, the distance between the centres of adjacent slits, is here the **'grating repeat'**.

Using Eqn (2.05) for R_1 as before, we obtain

$$R_N = R_1(0) \left(\frac{\sin \pi ua}{\pi ua}\right) \left(\frac{\sin N\pi uD}{\sin \pi uD}\right) \qquad (2.15)$$

The intensity is given by the square of this expression, viz.

$$I_N = I_1(0) \left(\frac{\sin \pi ua}{\pi ua}\right)^2 \left(\frac{\sin N\pi uD}{\sin \pi uD}\right)^2 \qquad (2.16)$$

Again, we have the (squared) product of the single-slit term and what is here the **'grating term'** – corresponding to the $\cos^2 \pi uD$ term in Eqn (2.13) when $N = 2$.

Figure 2.11 represents a graphical construction of the grating term. The fringes comprising the main peaks are the **principal maxima**. They occur when there is full reinforcement between the diffracted light from successive slits and, as for the double slit, this happens when $D\sin\theta = n\lambda$, i.e. $u = n/D$, where n is the **diffraction order** (cf. Eqn (2.10)).

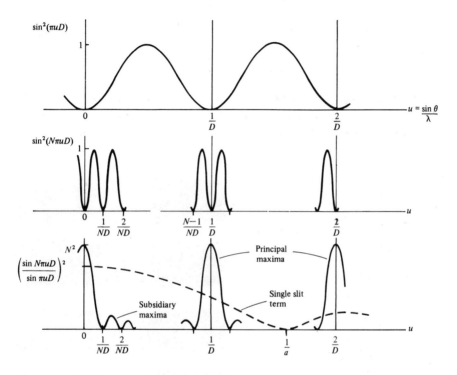

Fig. 2.11 – N-slit diffraction pattern.

In the directions of the principal maxima the value of the grating term is N^2, because as $u \to n/D$ we have

$$\left(\frac{\sin N\pi u D}{\sin \pi u D}\right)^2 \to \left(\frac{N\pi u D}{\pi u D}\right)^2 = N^2.$$

Since N is large for the type of grating normally used — usually tens of thousands — the principal maxima and the weaker, **subsidiary** (secondary) **maxima** become very sharp, as can be appreciated in Fig. 2.11. The figure also shows that when N is large the subsidiary maxima are close together. The overall effect on the subsidiary maxima is that they become insignificant in the pattern; we shall ignore them hereafter.

The intensity of illumination in the directions of the principal maxima is given by the square of the product of N and the amplitude of diffraction from each slit in those directions. For those directions Eqn (2.16) becomes

$$(I_N)_{\max} = I_1(0) \left[\frac{\sin \dfrac{\pi n a}{D}}{\dfrac{\pi n a}{D}} N \right]^2 \qquad (2.17)$$

which is the amplified sampled sinc^2 function, as for the double slit ($N = 2$). Figure 2.12 shows the patterns given by gratings with N up to 6. The way in which principal maxima are a sampling of the single-slit pattern is demonstrated.

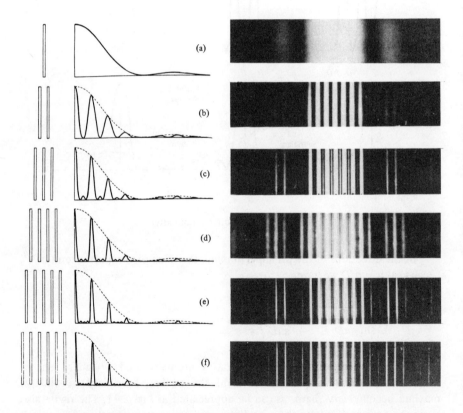

Fig. 2.12 – Diffraction patterns given by various numbers of slits.
(Reproduced from *Studies in Optics* by A. A. Michelson by permission of the University of Chicago Press. © 1927 by the University of Chicago.)

[§2.6]   2-Dimensional Gratings   43

The single-slit term and the grating term may both be described as existing in reciprocal space (§2.2). Indeed the spatial distribution of the principal maxima in the Fraunhofer diffraction plane can be described as the reciprocal lattice of the grating — albeit a 1-dimensional lattice in the present instance. Thus we can say that for large N, when the principal maxima are sharp, the pattern can be regarded as an amplified sampling of the single-slit pattern at the reciprocal-lattice points of the grating. This is an important concept.

Finally, it should be noted that as the pattern is symmetrical about the central (zero order) maximum, all the other maxima (meaning now the principal maxima) are symmetrically spaced in pairs, of diffraction order $\pm n$, about the central maximum.

2.6 2-DIMENSIONAL GRATINGS

No new principles are involved in extending the findings of the previous sections to two dimensions. The only result that we need to note here is summarized diagrammatically in Fig. 2.13. An aperture mask containing a large number of pinholes arranged to form a 2-dimensional lattice, of which just a small part is represented in (a), gives a diffraction pattern consisting of a lattice of spots arranged as indicated in (b). Consistent with previous results, the scale of the diffraction pattern is reciprocally related to that of the pinhole array and rows

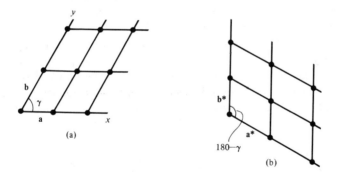

Fig. 2.13 — 2-dimensional diffraction.

of spots in the former are perpendicular to rows of pinholes in the latter. If **a,b** are the translation vectors of the grating lattice in (a), and **a*,b*** those of the diffraction pattern in (b) we have

$$|\mathbf{a}^*| = K/|\mathbf{a}|, \quad |\mathbf{b}^*| = K/|\mathbf{b}|$$

where K is a constant. Also we have

$$\mathbf{a}^* \cdot \mathbf{b} = 0, \quad \mathbf{b}^* \cdot \mathbf{a} = 0 \ .$$

In keeping with the reciprocity, a mask with pinholes arranged like the spots in (b) would give a diffraction pattern like the arrangement in (a). (It is a convention to use * to denote reciprocal lattice dimensions and it does not refer to the complex conjugate here.)

2.7 CRYSTALS AS 3-DIMENSIONAL GRATINGS

In crystals a group of atoms (sometimes — as in some metals — just one atom) is associated identically with each point of a regular space-lattice. The lattice repeats are generally between about 1 nm and 10 nm, though in crystals of biological molecules, for example, they are larger than this because so many atoms are associated with each lattice point. Since X-rays have wavelengths in the range from about 10^{-2} nm to about 10 nm (Appendix D) and are scattered (or diffracted — the terms are used interchangeably as noted in §1.3) by the electron clouds in atoms, crystals act like 3-dimensional diffraction gratings to X-rays. The 'Laue experiment' in 1912 demonstrated the reality of this, suggesting that contrary to some views at that time, X-rays are waves, or at least have wavelike properties. In 1914 Max von Laue was awarded the Nobel Prize in physics for this. In the following year the prize was awarded to W. L. Bragg and his father W. H. Bragg for their immediate application of the diffraction effects to the analysis of crystal structure by means of X-rays — now the subject of **X-ray crystallography**. As we have already noted, atoms cannot be imaged directly by focusing X-rays and the process has to be completed indirectly from the information provided by the diffraction effects.

Because a crystal is like a 3-dimensional grating, not just 1- or 2-dimensional, the conditions necessary for the equivalent of optical principal diffraction maxima to occur are not easily fulfilled. Consider the unit cell of a crystal lattice as depicted in Fig. 2.14(a). Imagine a train of quasimonochromatic waves of wavelength λ to sweep through the crystal. What is the basic requirement for

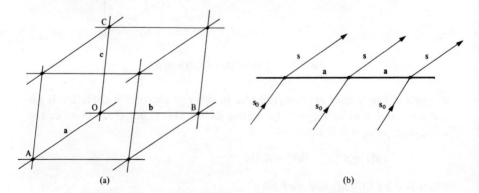

Fig. 2.14 — A crystal lattice as a 3-dimensional diffraction grating.

obtaining a diffraction maximum in some direction? It is that the X-ray scattering in that direction by the (identical) atomic assemblages centred at lattice points A, B, and C must be in phase with that scattered from the assemblage at O. The waves scattered from these centres will then be in phase with the scattering from their neighbours, and so on throughout the crystal. The fact that not even one atom need be located actually at a lattice point does not affect this requirement for a diffraction maximum because the lattice repeat — the distance between corresponding atoms located identically in relation to successive lattice points — is all that matters. The lattice point is of course hypothetical, and its location with respect to the atomic assemblage that it represents is arbitrary, though other considerations, such as symmetry, influence its choice. On the other hand, the *intensity* of a diffraction maximum is affected by the mutual positions of the atoms forming the assemblage, and we return to this later in this section.

Consider the repetition of lattice points along the a-direction in the crystal (Fig. 2.14(b)). For X-rays incident in a direction specified by unit vector s_0 the nett scattering from the assemblage at one lattice point in a direction specified by unit vector s, will be in phase with that from the next if

$$\mathbf{a} \cdot \mathbf{s} - \mathbf{a} \cdot \mathbf{s}_0 = (\mathbf{s} - \mathbf{s}_0) \cdot \mathbf{a} = h\lambda$$

where h is an integer or zero and $\mathbf{a}$ is the lattice translation vector. Neither s nor s_0 need be in the plane of the figure.

For the complete 3-dimensional array three conditions must simultaneously be fulfilled, viz:

$$\left. \begin{array}{l} (\mathbf{s} - \mathbf{s}_0) \cdot \mathbf{a} = h\lambda \\ (\mathbf{s} - \mathbf{s}_0) \cdot \mathbf{b} = k\lambda \\ (\mathbf{s} - \mathbf{s}_0) \cdot \mathbf{c} = l\lambda \end{array} \right\} \quad \text{Laue equations} \qquad (2.18)$$

where h,k,l are integers or zero, and $\mathbf{a},\mathbf{b},\mathbf{c}$ are the translation vectors of the lattice. Note that the lattice need not be orthogonal.

With a crystal at a fixed orientation to the direction of an incident X-ray beam it is unlikely that the above conditions can be fulfilled for more than a few diffraction maxima unless — as in the original experiment — a continuous spectrum of X-ray wavelengths is used. Not knowing what wavelength is responsible for any particular diffraction maximum has obvious disadvantages, however. Except for certain types of investigation the common practice now is to use quasimonochromatic radiation and to change the inclination of the crystal systematically during exposure to the X-ray beam to enable the Laue conditions to be met. Details of how this is done need not concern us here. Suffice it to say that it is possible to obtain the 3-dimensional diffraction pattern given by a crystal. It forms a 3-dimensional lattice that is reciprocal to that of the crystal, as with the 2-dimensional gratings considered in the previous section.

The *directions* of the X-ray diffraction maxima from a crystal are determined, for a given wavelength of X-rays, by the dimensions of the crystal grating. By measuring those directions the lattice geometry can therefore be determined. Similarly, the *intensities* of the diffraction maxima are determined by the intensity of scattering in the directions of the maxima, from the atomic group associated identically with each lattice point in the crystal. Those intensities therefore contain information about the composition and arrangement of the atoms in the group associated with each lattice point.

All this is directly analogous to the way in which the optical diffraction-grating maxima are an amplified sampling of the diffraction from one grating aperture alone. And the pattern from one aperture is determined by its aperture function — the optical structure of the aperture.

We are indeed fortunate to be able to gather this amplified sampling of the scattering of X-rays from the unit cell of a crystal. Even if one unit cell of a crystal could be isolated and handled, its scattering of X-rays would be far too weak to measure. Instead, we have the enormously amplified signals from a real crystal. The price we pay is that these signals are restricted to certain directions determined by the crystal lattice geometry. However, the number of directions is adequate, and it is possible to construct a very detailed picture of the atomic arrangement and electron density distribution in a crystal structure. (The diffraction of X-rays by non-crystalline materials such as glasses and liquids also reveals information about their structure, but details of this would take us beyond the scope of this book.)

In Chapter 5 we shall see how the use of X-ray diffraction data to determine the arrangement of atoms in crystals is, in essence, an application of the Abbe–Porter theory of optical image formation. The close analogy between the principles involved in the imaging of crystal structure by X-rays and conventional optical imaging was a topic of almost life-long interest to Sir Lawrence Bragg, a topic to which he made important contributions over many years.

In connection with the Fourier aspects that we are particularly concerned with it will be useful to derive an expression for the amplitude of an X-ray diffraction maximum from the unit cell of a crystal. This can be done by using the Bragg reflection concept, with which the reader should be familiar. (It is shown in Appendix C that a diffraction maximum corresponding to a particular set of values of h,k,l in the Laue equations may be regarded as a reflection of the incident X-ray beam by lattice planes in the crystal, defined by the same h,k,l values.)

Figure 2.15 represents an edge-on view of parts of a set of lattice planes of spacing d. X-rays are incident to the planes at the Bragg angle, θ, for the X-ray wavelength used. Thinking in terms of 'point atoms' for the moment, atoms in any one lattice plane scatter in phase in direction θ: this is the essence of 'reflection'. For reflections from successive lattice planes to be in phase the

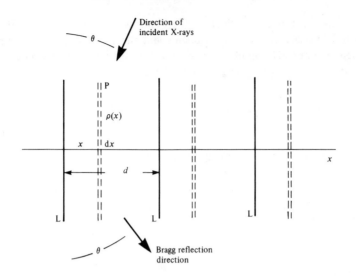

Fig. 2.15 — Bragg reflection.
L: edge-on view of lattice plane.

requirement of the Bragg equation has to be met, namely

$$2d \sin \theta = n\lambda \tag{2.19}$$

where n is the order of reflection.

As we have already noted, however, atoms are not point scatterers, nor are they confined to lattice points. It is necessary to think in terms of a continuous distribution of scattering matter, that repeats regularly throughout a crystal. (The symmetry aspects of a crystal structure need not distract us here.) Consider, with the aid of Fig. 2.15, the first-order reflection ($n = 1$) associated with the lattice spacing d. The scattering from the part of the electron-density distribution actually in the lattice planes is all in phase in the direction of the reflection, as we have noted above: but we must take into account the scattering from the matter between these planes. For the first-order reflection there is a path difference of one wavelength (a phase difference of 2π) in the scattering from successive lattice-planes. The contribution to the same reflection from the scattering matter in plane P is therefore out of phase by $(x/d)2\pi$. If $\rho(x)$ is the amplitude of scattering from a layer of thickness dx at x, then the total amplitude of this particular reflection from all the scattering matter in one lattice repeat of the structure is given by

$$A_1(\theta) = \int_0^d \rho(x)\, e^{i(x/d)2\pi}\, dx \ . \tag{2.20}$$

For the second-order reflection there are two wavelengths path difference in the reflections from successive planes and the phase difference for plane P will then be $(2x/d)2\pi$. Similarly, for the nth order reflection the phase will be $(nx/d)2\pi$ and the corresponding equation becomes

$$A_n(\theta) = \int_0^d \rho(x)\, e^{i(nx/d)2\pi}\, dx \; , \tag{2.21}$$

where θ is related to d by the Bragg equation (2.19).

'Reflection' is of course possible at each side of a set of planes and the amplitudes are usually equal (Friedel's Law). One can therefore regard this equation as representing the amplitude of pairs of reflections, or equally, the amplitude of pairs of diffraction maxima, of order $\pm n$.

3

Fourier series and periodic structures

> Fourier's theorem is not only one of the most beautiful results of modern analysis, but it may be said to furnish an indispensable instrument in the treatment of nearly every recondite question in modern physics.
>
> *attrib. Lord Kelvin*

3.1 INTRODUCTION

In Chapter 2 we have seen, in a rather restricted way, how diffraction data given by a periodic object in the form of an optical grating are determined by the 'structure' of the grating as specified by its aperture function; and that the same is true of the X-ray diffraction data obtained from the atomic arrangement comprising the periodic grating-like structure of a crystal. We have also noted that optical diffraction is an intermediate step when images are formed with a lens, the lens performing the task of recombing the diffracted light, in the image plane. With X-rays the action of a lens is not available, and to form an image of the structural arrangement of atoms in a crystal with X-rays other means have to be used to complete the imaging process beyond the diffraction stage.

For a deeper understanding of these matters, and indeed for other reasons that are described in Chatper 5, we need to know more about the relationship between a diffraction pattern and the structure of the object giving it. This is where Fourier methods start to play their vital role.

Continuing, for the time being, to restrict ourselves to periodic objects, such as multiple-aperture gratings or crystals, the overall aperture function (or 'structure') of these can be built-up mathematically by the summation of an infinite series of sinusoidal harmonics – a **Fourier series**, named after *J. B. J. Fourier,* the pioneer of the mathematical technique. This statement that a periodic function can be expressed as the sum of a series of harmonics of the function, is known as **Fourier's theorem**. Most functions obey it, but details of exceptions – though none are encountered in this book – will be found in mathematics textbooks.

Strictly, the pattern that can be represented by a Fourier series should be repeated an infinite number of times. However, for the applications we shall meet the number of repeats will be sufficient for the use of the series to be justified. In most practical gratings the number of repeats is very large, and in a crystal of 1 mm linear dimension there are more than a million repetitions of the atomic 'pattern unit'.

In this chapter we shall witness the extraordinary fact that the diffraction pattern given by a grating is a statement, almost a physical manifestation, of those very harmonics that comprise a mathematical description of the structure of the grating.

Since diffraction is an intermediate step in optical image formation, one consequence is that at the diffraction stage we can control in a calculated way the formation of an image; this is the basis of many aspects of optical processing (Chapter 5). Another major consequence has been the development of methods for the determination of the atomic structure in crystals, no matter how complex, from their X-ray diffraction data.

In Chapter 4 the relationships concerning periodic structures are extended to ones that are non-periodic. But first, some details about Fourier series (§3.2 and §3.3) and their relationship to diffraction by periodic structures (§3.4).

3.2 FOURIER SERIES

The use of Fourier series for representing a periodic pattern such as the optical structure of a grating, can be illustrated by considering the pattern, of which only two units are shown, in Fig. 3.01(a). This particular shape is chosen because it shows rather clearly how it can be represented by a Fourier series, and also because of the way it can represent an interesting example later (§3.4.2).

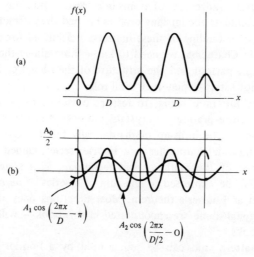

Fig. 3.01 — Fourier series representation of a periodic function $f(x)$.

Fourier Series

A Fourier series to represent a 1-dimensional periodic function, $f(x)$, can be written

$$f(x) = \frac{A_0}{2} + A_1 \cos\left(\frac{2\pi x}{D} - \alpha_1\right) + A_2 \cos\left(\frac{2\pi x}{D/2} - \alpha_2\right) + \ldots$$

$$= \frac{A_0}{2} + \sum_{n=1}^{\infty} A_n \cos\left(\frac{2\pi n x}{D} - \alpha_n\right) \quad (3.01)$$

i.e. an infinite series of harmonic 'waves' with amplitudes A_n (the Fourier **coefficients**), phases α_n, and progressively decreasing 'wavelengths' D/n — hereafter called 'repeats' to avoid confusion with optical wavelengths. $A_0/2$ is a constant, the factor $1/2$ serving only to simplify the writing of some equations later.

An analytical procedure for determining the coefficients for any given $f(x)$ is described in §3.3. Meanwhile, Fig. 3.01(b) illustrates graphically how the first few terms in the **Fourier analysis** of the pattern can at least be recognized qualitatively. In this figure we see that the term for $n = 1$, i.e. $A_1 \cos((2\pi x/D) - \alpha_1)$, has some similarity to $f(x)$ if $\alpha_1 = \pi$. The value of α_1 can only be 0 or π because $f(x)$ in this example is an 'even' function and $\alpha = 0$ would make the resemblance worse. At least the high peaks in the pattern are roughly represented but not the weak ones. The term for $n = 2$, with $\alpha_2 = 0$, helps considerably. Added to the $n = 1$ term with a suitable value of its amplitude A_2, it starts to establish the weak peaks and to enhance the strong ones. Addition of higher harmonics progressively sharpens the match. So far, however, the summation is equally distributed above and below the x-axis. It is the constant term $A_0/2$ in Eqn (3.01) that raises the pattern up from the x-axis to the actual level of the $f(x)$ curve.

The α_n in Eqn (3.01) are the 'phases' of the harmonics. For example, phase $\alpha_1 = \pi$ means that the cosine term for $n = 1$ is shifted along the x-axis by $D/2$ compared to a cosine of zero phase.

The summation of the harmonics, the **Fourier synthesis**, is formally to $n = \infty$. In practice, however, fewer than a hundred terms are often found to be quite adequate, though of course in some instances many more are needed.

In a similar way to that for the pattern we have just considered, the structure of an optical diffraction grating can be represented by a Fourier series. For example, as an extension of the notion of an 'aperture function' of a single slit (§2.2) the aperture function of a uniformly illuminated multiple-slit grating can be represented by the function $f(x)$ in Fig. 3.03(a). The Fourier series to represent this, and the extremely important relationship between the series and the diffraction pattern given by the grating, are dealt with in §3.4. To that end, the method used for the quantitative determination of the terms in a Fourier series is established in the next section.

In connection with our use of Fourier series in diffraction, it is particularly helpful to think of the frequencies of the harmonics as well as their repeats. Just as a temporal periodicity T has a temporal frequency $1/T$, so does a spatial harmonic of repeat D/n have a **spatial frequency** of n/D; it is the number of repeats in one unit of pattern.

3.3 DETERMINING FOURIER COEFFICIENTS: EVEN FUNCTIONS

For the even functions considered in the previous section, Eqn (3.01) can be written as

$$f(x) = \frac{A_0}{2} + \sum_{n=1}^{\infty} A_n \cos \frac{2\pi nx}{D} , \qquad (3.02)$$

if it is understood that A_n is not necessarily positive.

Multiplying both sides by $\cos(2\pi hx/D)$, where $h = 0$ or an integer, and integrating over the repeat D, we have

$$\int_D f(x) \cos \frac{2\pi hx}{D} \, dx = \frac{A_0}{2} \int_D \cos \frac{2\pi hx}{D} \, dx + \int_D \sum_{n=1}^{\infty} A_n \cos \frac{2\pi nx}{D} \cos \frac{2\pi hx}{D} \, dx$$

$$= \frac{A_0}{2} \int_D \cos \frac{2\pi hx}{D} \, dx + \frac{1}{2} \left[\int_D \sum_{n=1}^{\infty} A_n \cos \frac{2\pi(n+h)x}{D} \, dx + \int_D \sum_{n=1}^{\infty} A_n \cos \frac{2\pi(n-h)x}{D} \, dx \right].$$

The integrand of every term in the first summation on the right side of this equation is zero, since n is an integer and $h = 0$ or an integer. The same applies to the second summation except for the one term where $h = n$, which is $\int_D A_n \cos 0 \, dx = A_n D$. The first term on the right side is also zero for $h = n \neq 0$, so the whole equation reduces to

$$\int_D f(x) \cos \frac{2\pi nx}{D} \, dx = \frac{1}{2} \left[0 + A_n D \right]$$

whence $\qquad A_{n \neq 0} = \dfrac{2}{D} \int_D f(x) \cos \dfrac{2\pi nx}{D} \, dx$.

When $h = 0$ only the first term is non-zero, and this gives

$$\int_D f(x) \, dx = \frac{1}{2} A_0 D$$

[§3.3] Determining Fourier Coefficients: Even Functions

whence $\quad A_0 = \dfrac{2}{D}\displaystyle\int_D f(x)\,dx$.

The reason for the arbitrary factor of $\tfrac{1}{2}$ in the constant term in Eqn (3.01) is now evident: it enables the above two expressions for the coefficients to be combined as

$$A_n = \frac{2}{D}\int_D f(x)\cos\frac{2\pi n x}{D}\,dx \qquad (3.03)$$

where $n = 0,1,2,\ldots$.

The expression for the constant term, $A_0/2$, is simply the average value of $f(x)$ over the whole pattern-unit, of repeat D. This confirms and quantifies the observation made in §3.2 that $A_0/2$ is a constant amount by which the summation has to be raised from the x-axis to match the height of the profile of $f(x)$.

For $n \neq 0$, the interpretation of Eqn (3.03) is fascinating. As illustrated graphically in Fig. 3.02 (though normally it would be done by computation) it means that if we want to know whether a harmonic of a certain spatial frequency n/D is present in the Fourier series representation of a given function,

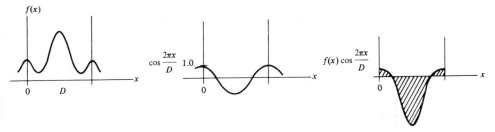

Fig. 3.02 – Graphical construction of $A_1 = \displaystyle\int_D f(x)\cos(2\pi x/D)\,dx$.

we multiply the pattern-unit of the function by a harmonic of that frequency (unit amplitude and zero phase) and then integrate over D. A zero answer indicates that the harmonic is not a constituent of the function. A non-zero answer indicates that it is a constituent, and the value obtained by the integration (i.e. by the area under the product curve), after multiplication by $2/D$ as required by Eqn (3.03), gives its amplitude. The phase of the harmonic is revealed by the sign of the value obtained – whether the nett area under the product curve is positive or negative.

For our purposes there is an advantage in rewriting Eqn (3.02) in the form

$$f(x) = \sum_{n=-\infty}^{+\infty} \frac{A_n}{2}\cos\frac{2\pi n x}{D} \quad . \qquad (3.04)$$

The term for $n = 0$ is the same as before. For other values of n the coefficients are now in pairs, $\frac{1}{2}A_{+n}$ and $\frac{1}{2}A_{-n}$, which are equal ($\cos(-x) = \cos(+x)$) and have the same spatial frequencies. They therefore add to give the same harmonic amplitude as when only positive values of n were involved. It follows that Eqn (3.03) now gives the coefficients $\frac{1}{2}A_n$ in Eqn (3.04) for all n, viz.

$$\frac{A_n}{2} = \frac{1}{D} \int_D f(x) \cos \frac{2\pi n x}{D} \, dx \, . \tag{3.05}$$

The reason for modifying the series to consist of pairs of harmonics of half the original amplitude, but for negative as well as positive values of n, is because these pairs are seen in the following section to have a physical counterpart in the pairs of diffraction maxima of order $\pm n$ given by a grating.

General formulae for $f(x)$ when this is not an even function are given in §3.5.

3.4 OPTICAL AND CRYSTAL DIFFRACTION GRATINGS: PHYSICAL INTERPRETATION OF FOURIER TERMS

3.4.1 Optical diffraction

The grating whose aperture function, or optical structure as we are also describing it, is represented by $f(x)$ in Fig. 3.03(a) has slits of width a and grating repeat D. The (uniform) amplitude of illumination at the slits is h.

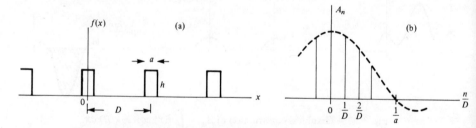

Fig. 3.03 — Fourier analysis of a periodic rectangle-function.

The Fourier analysis of $f(x)$ is given by substituting these details into Eqn (3.05) as follows:

$$\frac{A_n}{2} = \frac{1}{D} \left[\int_0^{a/2} h \cos \frac{2\pi n x}{D} \, dx + \int_{D-a/2}^{D} h \cos \frac{2\pi n x}{D} \, dx \right]$$

whence

$$\frac{A_n}{2} = \frac{ha}{D} \left[\frac{\sin \frac{\pi n a}{D}}{\frac{\pi n a}{D}} \right] . \tag{3.06}$$

Plotted in Fig. 3.03(b), they give the amplitudes and phases of the pairs ($\pm n$) of harmonics of repeat D/n (spatial frequency n/D) comprising $f(x)$. Graphically, and in Eqn (3.06), they are also immediately recognizable as the **sampled sinc function** that tells us the amplitudes and phases of the pairs of diffraction maxima, or order $\pm n$, given by a grating whose overall aperture function is $f(x)$ (cf. the intensity values in Eqn (2.17)). And the directions of those diffraction maxima are given by the spatial frequencies of the Fourier harmonics because

$$\frac{n}{D} = \frac{\sin\theta}{\lambda} (= u) \quad . \qquad \{2.10\}$$

The above findings are applicable to practically any type of aperture function, though of course the envelope of the coefficients is not generally a sinc function.

We have the important general result then, that the *amplitude* of the nth order pair of diffraction maxima from a grating is a measure of the amplitude of the nth order pair of harmonics comprising its overall aperture function — what we have called its optical structure. The *direction* of those diffraction maxima gives, as $(\sin\theta)/\lambda$, the spatial frequency (n/D) of the harmonics (or the harmonic repeat, D/n).

For these reasons the plane of a diffraction pattern is referred to as the **Fourier plane** (or Fourier space), or alternatively as the (spatial) **frequency plane** (or domain). Furthermore, as first noted in Chapter 2, the term *reciprocal space* can also be used, because of the reciprocal relationship between the scale of a diffracting system and the pattern it gives. Each of these interpretations has its particular relevance and usefulness, e.g. the 'frequency domain' is widely used in connection with optical data processing (§5.5).

3.4.2 Crystal diffraction

In §2.7 it was shown that the amplitude, $A_n(\theta)$, of a Bragg reflection of order n, at crystal lattice planes of spacing d, is given by

$$A_n(\theta) = \int_{x=0}^{x=d} \rho(x)\, e^{i(nx/d)2\pi}\, dx \qquad \{2.21\}$$

where θ is related to d by the Bragg equation.

Let us apply this to the crystal structure of caesium chloride, CsCl (Fig. 3.04). The electron density distribution through the structure in the direction of one of the cube edges can be represented by the curve $f(x)$ used in Fig. 3.01(a); it was deliberately chosen for this purpose. The low peaks in that pattern represent the electron cloud of the chloride ions and the high peaks that of the caesium ions. (There are equal numbers of each in the structure.)

If we ignore various angle-dependent (and other) factors, we can assume for the present purpose that the amplitude of X-ray scattering $\rho(x)$ in the above

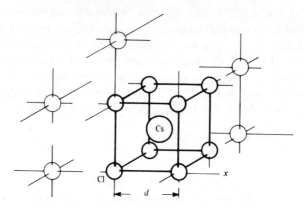

Fig. 3.04 — The crystal structure of caesium chloride, CsCl.

equation is equal to $f(x)$, and since this particular $f(x)$ is an even function we can rewrite the above equation as

$$A_n(\theta) = \int_{x=0}^{x=d} f(x) \cos \frac{2\pi n x}{d} \, dx \, . \tag{3.07}$$

Comparison with Eqn (3.05) shows that these values for the diffraction amplitudes are proportional to the amplitudes of the harmonic terms of the Fourier series analysis of the electron density distribution between the reflecting planes in the crystal structure.

The same relationship applies to the full 3-dimensional nature of crystal structures and it is readily extended to deal with examples that are not centrosymmetric. Thus we have the fundamental and outstandingly important result that measurement of X-ray diffraction maxima — more conveniently regarded as Bragg reflections — provides information about the terms in the Fourier series analysis of the crystal structure, just as with the 'structure' of an optical diffraction grating. An experimental snag is that because only the intensities of the diffracted X-rays can be measured the phases of the Fourier terms are unknown and a Fourier synthesis cannot be performed directly (§5.3.3).

The notes at the end of the previous section concerning the ways of describing the space in which a diffraction pattern is formed are equally applicable here. The concept of reciprocal space is particularly useful in X-ray crystallography.

3.5 FOURIER SERIES: GENERAL FORMULATIONS

When $f(x)$ is a general periodic function the α_n in Eqn (3.01) can have values from 0 to 2π. The following two alternative forms of the series are then often more useful.

3.5.1 The sine and cosine series
Expand Eqn (3.01) as follows:

$$f(x) = \frac{A_0}{2} + \sum_{n=1}^{\infty}\left[A_n \cos \alpha_n \cdot \cos \frac{2\pi n x}{D} + A_n \sin \alpha_n \cdot \sin \frac{2\pi n x}{D}\right]$$

This can be written as

$$f(x) = \frac{A_0}{2} + \sum_{n=1}^{\infty} a_n \cos \frac{2\pi n x}{D} + \sum_{n=1}^{\infty} b_n \sin \frac{2\pi n x}{D} \tag{3.08}$$

where $a_n = A_n \cos \alpha_n$, $b_n = A_n \sin \alpha_n$.

The original cosine series with various phases has been replaced by a combination of cosines and sines

(cf. $2 \cos(2\pi n x/D - \pi/3) = 1 \cos 2\pi n x/D + \sqrt{3} \sin 2\pi n x/D$).

The following expressions for the new coefficients are obtained in a similar way to that used for the single series earlier:

$$\left.\begin{aligned} A_0 &= \frac{2}{D} \int_D f(x)\, dx \\ a_n &= \frac{2}{D} \int_D f(x) \cos \frac{2\pi n x}{D}\, dx \\ b_n &= \frac{2}{D} \int_D f(x) \sin \frac{2\pi n x}{D}\, dx \end{aligned}\right\} \tag{3.09}$$

The reader can check that for an even function all $b_n = 0$ and that, similarly, for an odd function all $a_n = 0$.

3.5.2 Exponential notation
For $f(x)$ real we can use Eqn 3.01 and rewrite it as

$$f(x) = \frac{A_0}{2} + \sum_{n=1}^{\infty} \frac{1}{2} A_n e^{i\left[\frac{2\pi n x}{D} - \alpha_n\right]} + \sum_{n=1}^{\infty} \frac{1}{2} A_n e^{-i\left[\frac{2\pi n x}{D} - \alpha_n\right]}.$$

The negative sign in front of i in the second summation may be combined with the n, making it a summation over negative values of n. The equation can therefore be changed to

$$f(x) = \frac{A_0}{2} + \sum_{n=1}^{\infty} \frac{1}{2} A_n e^{i\left[\frac{2\pi nx}{D} - \alpha_n\right]} + \sum_{n=-1}^{-\infty} \frac{1}{2} A_n e^{i\left[\frac{2\pi nx}{D} + \alpha_n\right]}$$

which can be compressed to

$$f(x) = \sum_{n=-\infty}^{+\infty} C_n e^{\frac{i2\pi nx}{D}} \quad . \tag{3.10}$$

The complex coefficients C_n contain the A_n and α_n as follows

$$C_{n>0} = \frac{1}{2} A_n e^{-i\alpha_n}$$

$$C_{n<0} = \frac{1}{2} A_n e^{+i\alpha_n} \tag{3.11}$$

$$C_{n=0} = \frac{1}{2} A_0 \quad .$$

A consequence of this is that

$$\left. \begin{array}{l} C_{n<0} = C^*_{n>0} \\ \\ \text{and therefore } |C_{n<0}| = |C_{n>0}| \end{array} \right\} \tag{3.12}$$

where * denotes the complex conjugate.

The coefficients are derived in an analogous way to that for the cosine series and we have the pair of equations

$$f(x) = \sum_{n=-\infty}^{+\infty} C_n e^{\frac{i2\pi nx}{D}} \quad \{3.10\}$$

$$C_n = \frac{1}{D} \int_D f(x) \, e^{\frac{-i2\pi nx}{D}} \, dx \quad . \tag{3.13}$$

Fourier Series: General Formulations

As an example, consider the exponential Fourier series for the function $f(x) = \cos \frac{2\pi 5x}{D}$. Of course this expression could be substituted into Eqn (3.13) and the Fourier coefficients C_n obtained. However, the answer is obtainable by inspection because

$$\cos \frac{2\pi 5x}{D} = \frac{1}{2}\left[e^{\frac{i2\pi 5x}{D}} + e^{\frac{-i2\pi 5x}{D}}\right].$$

The exponential series for $f(x)$ consists of only these two terms, viz. terms for which $n = \pm 5$ with coefficients $1/2$.

Similarly, if $f(x) = \sin \frac{2\pi 5x}{D}$ only terms $n = \pm 5$ are non-zero, and the coefficients are $\pm \frac{1}{2} i$ respectively.

Figure 3.05 shows these two examples in graphical form.

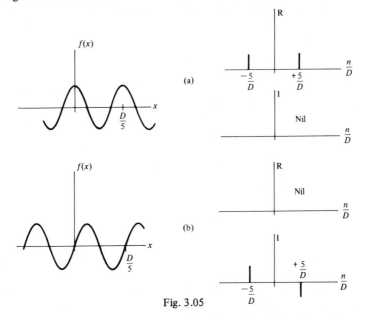

Fig. 3.05

Eqn (3.13) was derived from Eqn (3.01), i.e. for $f(x)$ real, and the presence of imaginary numbers only arises in the way that we have just seen, the whole expression being still real.

The above examples also illustrate the condition stated in Eqn (3.12). Thus for real functions that are even, the pairs of coefficients are real and of the same sign, whereas for real functions that are odd, the pairs are imaginary and of opposite sign.

For **complex functions** Eqn (3.12) does not apply. This is quickly illustrated by considering the function

$$\sin\frac{2\pi 5x}{D} + \frac{i}{2}\cos\frac{2\pi 5x}{D}$$

for which the coefficients are $\mathbf{C}_5 = -i/4$, $\mathbf{C}_{-5} = 3i/4$.

As a final example of the exponential Fourier series the reader should derive the coefficients for the repeating rectangle function in Fig. 3.03(a). It will be found that inserting the details of that function into Eqn (3.13) gives

$$C_n = \frac{ha}{D}\left[\frac{\sin\frac{\pi na}{D}}{\frac{\pi na}{D}}\right] \qquad (3.14)$$

Here is the sampled sinc function as before (Eqn (3.06)). Since the function is real and even the coefficients are real and equal in pairs ($\pm n$) as found in §3.3 using the cosine series.

3.5.3 Space and time

The Fourier series described in this chapter are readily extended to 2- and 3-dimensions. Also, when the patterns analysed are functions in time instead of space, the Fourier coefficients refer to temporal frequencies instead of spatial frequencies. The corresponding equations in exponential notation are then

$$f(t) = \sum_{n=-\infty}^{+\infty} C_n\, e^{\frac{i2\pi nt}{T}} \qquad (3.15)$$

$$C_n = \frac{1}{T}\int_T f(t)e^{\frac{-i2\pi nt}{T}}\,dt\ . \qquad (3.16)$$

For example, if the repeating rectangle function we have used represents a stream of pulses of duration a and repeat time T, then the Fourier terms form the frequency spectrum required for the generation of the pulses.

In §4.6 the application of Fourier methods to a single time-pulse leads to an interpretation of the frequency spectrum of the wavetrains associated with light photons.

4

Fourier transforms, convolution and correlation

4.1 INTRODUCTION

In Chapter 3 the harmonic terms comprising the Fourier analysis of the optical structure of an object in the form of a multiple-aperture grating have been identified with details of the diffraction pattern given by the grating. There is an analogous relationship between a single aperture and its diffraction pattern, which we need to explore. It is the **Fourier transform**, involving – as one would expect for a non-repeating pattern – an integral rather than a series. This is the subject of §4.2.

The Fourier transform also has other important roles in physical optics; indeed it would be difficult to exaggerate its importance generally in physics. This chapter is concerned with the further insight it provides into the relationship between the diffraction pattern given by a multiple-aperture diffracting system such as a grating or crystal, and its (overall) aperture function or structure. The basic ideas of this are introduced in §4.3–§4.5 for use in various ways in Chapter 5 in connection with image formation and image processing. In §4.3 we consider the grating diffraction pattern, and in §4.4 the aperture function of the grating. The latter is considered in terms of convolution – another concept and mathematical procedure widely used in physics. In §4.5 the two sides of the relationship, the aperture function of the grating and the diffraction pattern it gives, are brought together and seen as an example of the convolution theorem.

In §4.6 we consider something quite different, the role of the Fourier transform in relating the length of a wavetrain of light to its spectral composition. This is of importance when we turn our attention in Chapter 6 to the analysis of radiation sources by interferometry.

Finally in §4.7, we introduce correlation, an operation of considerable importance in both Chapters 5 and 6.

4.2 THE FOURIER TRANSFORM AND SINGLE-SLIT DIFFRACTION

First let us recall the repeating rectangle function (Fig. 4.01(a)). With the chosen

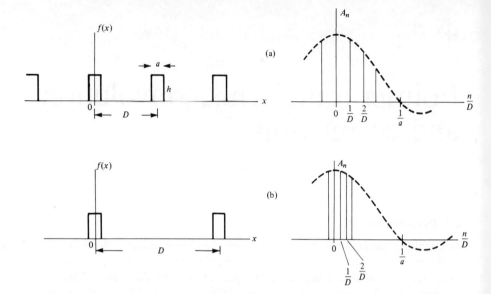

Fig. 4.01 – Fourier analysis of a periodic rectangle-function.

origin this is an even function, and in §3.4.1 it was noted that if it is represented by the Fourier series

$$f(x) = \sum_{n=-\infty}^{+\infty} \frac{A_n}{2} \cos \frac{2\pi n x}{D} \qquad \{3.04\} \ (4.01)$$

then the coefficients,

$$\frac{A_n}{2} = \frac{1}{D} \int_D f(x) \cos \frac{2\pi n x}{D} \, dx \qquad \{3.05\} \ (4.02)$$

become

$$\frac{A_n}{2} = \frac{ha}{D} \left[\frac{\sin \frac{\pi n a}{D}}{\frac{\pi n a}{D}} \right]. \qquad \{3.06\} \ (4.03)$$

This last expression contains what we have called a sampled sinc function and it is shown again in Fig. 4.01(a). Furthermore, if $f(x)$ represents the aperture function of a whole diffraction grating, of repeat D, we have identified it with the amplitude of light from each slit in just the directions ($\pm \theta$) of the grating

[§4.2]   The Fourier Transform and Single-Slit Diffraction

maxima (order $\pm n$). For it is the single-slit pattern, $(\sin \pi ua)/(\pi ua)$, in which u is restricted to specific values given by

$$u = \frac{\sin \theta}{\lambda} = \frac{n}{D} \qquad \{2.10\} \quad (4.04)$$

$D \sin \theta = n\lambda$ being the condition for a grating maximum (§2.5).

We should recall, too, that we also have an interpretation of the spatial frequencies, n/D, of the harmonics in the Fourier analysis (Eqn (4.03)) of the grating aperture function. They tell us, via Eqn (4.04), the directions ($\pm \theta$) of the principal diffraction maxima given by the grating.

Now consider what happens if D is increased, as illustrated in Fig. 4.01(b). The spatial frequencies, and therefore the directions of the diffracted beams, become closer together. This suggests that for the limiting case of a non-periodic function, all frequencies are allowed. That is certainly consistent with our knowledge of diffraction: with a single aperture the diffraction pattern is a continuous function of u.

In Eqns (4.01) and (4.02), n and D then lose their significance and we have the integrals

$$f(x) = \int_{-\infty}^{+\infty} F(u) \cos 2\pi ux \, du \qquad (4.05)$$

$$F(u) = \int_{-\infty}^{+\infty} f(x) \cos 2\pi ux \, dx \qquad (4.06)$$

In the example we have been considering, $f(x)$ is an even function and $F(u)$ is said to be the **cosine Fourier transform** of $f(x)$. Eqn (4.06) corresponds to Eqn (4.02) and the discrete summation in Eqn (4.01) is replaced by the integration in Eqn (4.05). Eqn (4.03) becomes

$$F(u) = ha \left(\frac{\sin \pi ua}{\pi ua} \right) \qquad (4.07)$$

in agreement with the phasor derivation of the diffraction pattern for a single slit (cf. Eqn (2.05) where R_1 was the total amplitude of illumination at the slit, corresponding here to ha).

When $f(x)$ is not necessarily even we could, just as in §3.5.1, consider the general case in terms of sine and cosine integrals. However, exponential notation is generally more convenient, and it has the advantage of also being applicable to complex as well as real functions. By referring to Eqns (3.10) and (3.13) we can immediately write

$$f(x) = \int_{-\infty}^{+\infty} F(u) e^{2\pi iux} \, du \qquad (4.08)$$

$$F(u) = \int_{-\infty}^{+\infty} f(x)e^{-2\pi i u x}\, dx \ . \tag{4.09}$$

(It is immaterial which of these equations contains the negative sign.)

At the end of this section we consider the application of this exponential formulation to the simple case of the rectangle function, and see that it gives the same result as we have already obtained; but first some general comments.

The product of x and u in these equations, and in (4.05) and (4.06), is dimensionless: x and u are **conjugate parameters**. And because of the considerable 'symmetry' within each of these pairs of equations, $f(x)$ and $F(u)$ are each described as the **Fourier transform** (or 'integral') of the other. They constitute what is called a Fourier transform pair. (Conventionally, a distinction is made between 'transform' and 'inverse transform'. This arises from the negative sign in one exponential but it has no physical significance for us here.)

To see how Eqns (4.08) and (4.09) are used for complex functions, let us assume that we have $f(x)$ as a complex function and that we wish to calculate $F(u)$. We can express $f(x)$ as the sum of its real and imaginary parts as follows:

$$f(x) = v(x) + iw(x) \tag{4.10}$$

Using Euler's relation ($e^{i\theta} = \cos\theta + i\sin\theta$), Eqn (4.09) then becomes

$$F(u) = \int_{-\infty}^{+\infty} v(x)\cos(2\pi u x)\, dx + \int_{-\infty}^{+\infty} w(x)\sin(2\pi u x)\, dx$$
$$+ i\left[\int_{-\infty}^{+\infty} w(x)\cos(2\pi u x)\, dx - \int_{-\infty}^{+\infty} v(x)\sin(2\pi u x)\, dx\right]. \tag{4.11}$$

Each integrand is real and we see that $F(u)$ can be calculated in the usual way by combining the first two with the second two in the complex plane (Argand diagram).

Before continuing, we will just note how the exponential form reduces to simple forms when $f(x)$ is real, and to Eqn (4.05) in particular, when it is also even. To do this we put

where
$$F(u) = A(u) + iB(u)$$
$$A(u) = F(u)\cos\alpha(u)$$
$$B(u) = F(u)\sin\alpha(u)$$

and
$$\alpha(u) = \tan^{-1}\frac{B(u)}{A(u)}\ . \tag{4.12}$$

[§4.2] The Fourier Transform and Single-Slit Diffraction

Then we can write Eqn (4.08) as

$$f(x) = \int_{-\infty}^{+\infty} (A(u) + iB(u))[\cos(2\pi ux) + i\sin(2\pi ux)]\,du$$

$$= \int_{-\infty}^{+\infty} [A(u)\cos(2\pi ux) - B(u)\sin(2\pi ux)]\,du$$

$$+ i\int_{-\infty}^{+\infty} [B(u)\cos(2\pi ux) + A(u)\sin(2\pi ux)]\,du \quad . \quad (4.13)$$

If $f(x)$ is real then the imaginary term must be zero. This requires that

$$B(-u) = -B(+u)$$
$$A(-u) = A(+u)$$

which means that

$$F(u) = A(u) + iB(u), \quad u > 0$$

but

$$F(u) = A(u) - iB(u), \quad u < 0$$

i.e.

$$\left. \begin{array}{c} F(u) = F^*(-u) \\ \text{and therefore } |F(u)| = |F(-u)|. \end{array} \right\} \quad (4.14)$$

Thus for $f(x)$ real we have simply the first term in Eqn (4.13) i.e.

$$f(x) = \int_{-\infty}^{+\infty} A(u)\cos(2\pi ux)\,du - \int_{-\infty}^{+\infty} B(u)\sin(2\pi ux)\,du \quad (4.15)$$

or, with the help of Eqns (4.12), we can express this as

$$\left. \begin{array}{c} f(x) = \displaystyle\int_{-\infty}^{+\infty} F(u)\cos[2\pi ux + \alpha(u)]\,du \\ \\ = 2\displaystyle\int_{0}^{\infty} F(u)\cos[2\pi ux + \alpha(u)]\,du \end{array} \right\} \quad (4.16)$$

where $F(u)$ is restricted by Eqn (4.14). The first form of (4.16) reduces to Eqn (4.05) when $f(x)$ is even.

(The close similarity between the relationships given above and those referring to the use of exponential notation for Fourier series (§3.5.2) should be noted.)

As with Fourier series, transforms are equally applicable to 2- and 3-dimensions. They are also used with time and frequency as the conjugate pair (§4.6).

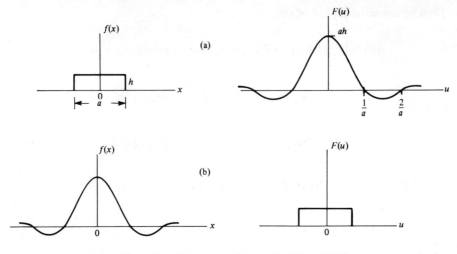

Fig. 4.02 — Fourier transform pairs, $f(x)$ and $F(u)$.

To return to the case of the single rectangle function (Fig. 4.02(a)), Eqn (4.09) gives the expected sinc function quite simply. We have

$$F(u) = \int_{-a/2}^{+a/2} h e^{-2\pi i u x} \, dx \qquad (4.17)$$

$$= ha \left(\frac{\sin \pi u a}{\pi u a} \right) \qquad (4.18)$$

in agreement with the previous derivations.

Note that Eqn (4.17) itself can be identified directly as the amplitude of diffracted light in a direction specified by u, from a uniformly illuminated slit of width a. This is evident from Fig. 4.03 where we see that the contribution from an element dx of wavefront at distance x from the origin (0) has an amplitude $h \, dx$ and a phase of $(2\pi/\lambda)x \sin \theta = 2\pi u x$, so that for the whole slit the nett amplitude is

$$\int_{-a/2}^{+a/2} h e^{2\pi i u x} \, dx \; .$$

The Fraunhofer diffraction pattern given by a single aperture in general can be said to be the Fourier transform of its aperture function. Note that the symmetry in the relationship between $f(x)$ and $F(u)$ is such that if $f(x)$ were an aperture function in the form of a sinc function, then its diffraction pattern, $F(u)$, would be a rectangle function. This exchange of functions is shown in Fig. 4.02(b).

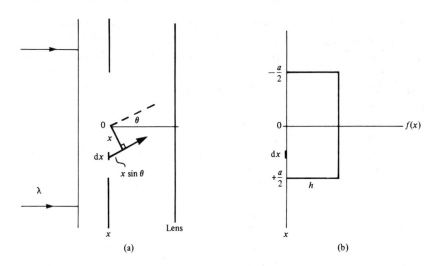

Fig. 4.03 – Single-slit Fraunhofer diffraction.

The earlier findings (§2.4, §2.5) concerning the sampling relationship between the diffraction pattern given by a grating and by a single slit may now be extended to describe the grating pattern as a sampling of the Fourier transform of the single-slit aperture function.

Both the Fourier transform of the single slit and the Fourier series of the grating are spatially specified by u, continuous in one and discrete in the other. Both may therefore be described as existing in *Fourier space*, or *frequency space*, as described in §3.4.1 in connection with the diffraction grating. This is a very useful generalization of the interpretation of diffraction, and is true for any aperture function.

The identification of the Fraunhofer diffraction pattern of an aperture function with the Fourier transform of that aperture function leads to the description of a lens as a **Fourier transformer**. We take lenses so much for granted that it is easy to overlook this remarkable property of what is after all a lump of glass.

4.3 THE GRATING PATTERN AS A PRODUCT OF TRANSFORMS

In the following expression for the diffraction pattern given by an N-slit grating

$$R_N = R_1(0)\left(\frac{\sin \pi u a}{\pi u a}\right)\left(\frac{\sin N\pi u D}{\sin \pi u D}\right) \qquad \{2.15\} \quad (4.19)$$

the 'single-slit' term, $(\sin \pi u a)/\pi u a$, has been seen in the previous section to be the Fourier transform of the aperture function of the slit.

Turning our attention now to the 'grating term' in this expression we shall find that it also is a Fourier transform, this time of the grating 'lattice' on which the individual slits are distributed. The result, that the diffraction pattern represented by Eqn (4.19) is therefore the product of two transforms, is of general validity for other forms of aperture function and for gratings of 1, 2, and 3 dimensions.

To establish the Fourier transform nature of the grating term the regular lattice on which the grating is built is specified by an array of 'markers' at each one of which is located an identical single aperture – here a slit. For such a marker we use the so-called δ-**function**, a mathematical device (not a true function in the mathematical sense) defined as the limiting form of a rectangle function (Fig. 4.02(a)) when, whilst maintaining constant area (usually chosen as unity), its width tends to zero while the height tends to infinity. The δ-function thus has a value of zero except at one specific point where it is infinite. In some contexts it is described as a (unit) **impulse function**. (No known ordinary function behaves like this and it is described as a generalized function: it is normally specified by the integral properties attached to it.)

First consider a single δ-function located at the origin of x in Fig. 4.04(a). It is denoted as the function $f(x) = \delta(x)$, and Eqns (4.17) and (4.18) show that its Fourier transform is unity (if $ha = 1$) for all u. This is what one would expect for the diffraction pattern from an 'ideally narrow' slit (a slit of infinitesimal width), as there would be no path differences like those causing interference in the light diffracted from a slit of finite width. In practice it would transmit an infinitesimal amount of light and it is in that sense that it is described above as a marker.

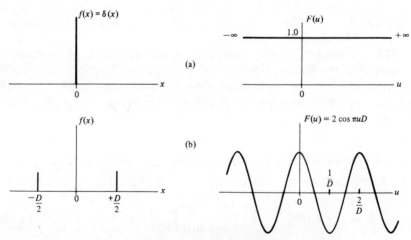

Fig. 4.04 – δ-functions and their Fourier transforms.
    (a) δ-function at the origin;
    (b) δ-functions at $x = \pm D/2$.

To build up an array of δ-functions to represent a grating it is necessary to specify δ-functions at other values of x. A δ-function moved from the origin to some position x_0 is expressed as

$$f(x) = \delta(x - x_0)$$

and its Fourier transform is

$$F(u) = \int_{-\infty}^{+\infty} \delta(x - x_0) e^{2\pi i u x} \, dx \ .$$

Put $x - x_0 = x'$, then $dx = dx'$ and we have

$$F(u) = \int_{-\infty}^{+\infty} \delta(x') e^{2\pi i u (x' + x_0)} \, dx'$$

$$= e^{2\pi i u x_0} \int_{-\infty}^{+\infty} \delta(x') e^{2\pi i u x'} \, dx'$$

$$= e^{2\pi i u x_0} \tag{4.20}$$

i.e. unit modulus with phase depending on x_0.

A multiple array in one dimension is similarly written as

$$f(x) = \sum_N \delta(x - x_N) \tag{4.21}$$

and its Fourier transform is

$$F(u) = \sum_N e^{2\pi i u x_N} \ . \tag{4.22}$$

As an example, consider two δ-functions at $x = \pm D/2$. The transform is then

$$F(u) = e^{\pi i u D} + e^{-\pi i u D}$$

$$= 2 \cos \pi u D \ .$$

This particular Fourier pair, $f(x)$ and $F(u)$, is illustrated in Fig. 4.04(b). $F(u)$ is readily interpretable as the diffraction pattern for two ideally narrow slits whose positions are specified by the δ-function markers defined by $f(x)$. It is the grating term in Eqn (4.19) for $N = 2$.

In the same way, the Fourier transform of a linear array of N equally spaced δ-functions is the grating term for N equally spaced slits.

We thus have the result that the two terms, the 'single-slit term' and the 'grating term', are respectively the Fourier transform of the single-aperture function and the transform of the array of δ-functions defining the grating lattice. That the diffraction pattern is the product of these two transforms is not restricted to the particular example chosen here.

4.4 CONVOLUTION

4.4.1 Introduction

In the previous section the aperture function of an entire grating was described as a distribution of its constituent single-aperture function in accordance with an array of δ-functions defining the lattice on which the grating is based. This distribution of one entity in a manner specified by another is an example of 'convolution'. It is a process that occurs in many forms and in numerous contexts. A Fourier aspect of convolution, embodied in the convolution theorem (§4.5), is of wide-ranging importance and usefulness — nowhere more so than in the areas of image formation and processing that are our concern here.

There are special features of convolution that are not particularly clearly illustrated by the example of the grating because of the discontinuous nature of δ-functions. As an introduction we will consider a more common type, the ubiquitous one in which convolution occurs when a continuous 'input' signal to a system is processed to give an 'output'.

Imagine that the function $f(x)$ in Fig. 4.05(a) represents a 1-dimensional input signal to some device. It could, for example, be the way the voltage to an amplifier varies with time: instead of $f(x)$ we would label it $f(t)$. Or it could be the spatial distribution of light intensity in a certain direction, x, across a screen. In the first of these the amplifier is probably required to increase (amplify) every ordinate in (a) by some factor, to form the output. In the second example, we may imagine that a photometer is faithfully to scan and measure the intensity distribution $f(x)$ and display it in some way. In both cases, however, the processing of the 'input' is usually less than perfect in practice. Every ordinate of the input becomes 'smeared' rather than cleanly reproduced as another ordinate in the output. Thus a δ-function input at $x = 0$ yields an output, $g(x)$ say, instead of a δ-function.

We shall assume that the 'shape' of the response of the system, including this unwanted smearing, is *invariant,* i.e. it is the same at every value of x. It is represented by the triangle in Fig. 4.05(b), shown at just two values of x, where it is $g(x)$ and $g(x - x_1)$ respectively, for unit strength of input. Consider what happens to the input at x_1 in (a). The input ordinate there is $f(x_1)$ and it is smeared by the response of the system to become the curve $f(x_1)g(x - x_1)$ shown by the full line in (c); thus $f(x_1)$ is the 'weighting' of the unit response $g(x - x_1)$. Only the ordinate at x_1 in this smearing actually contributes to the output at x_1. This output ordinate is not shown in the figure because it is necessary to see that

[§4.4] Convolution 71

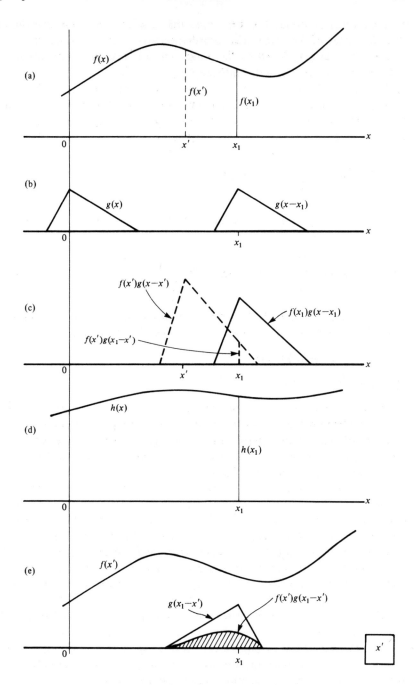

Fig. 4.05 – Convolution of $f(x)$ and $g(x)$.

there are other contributions there, ones that originate from the smearing of other ordinates in $f(x)$. One such contribution arises from the smearing of the ordinate at x' in (a), shown by the broken curve in (c). The contribution this makes to the output at x_1 is the ordinate $f(x')g(x_1-x')$.

In this way the total output at x_1, shown as h in (d), is the sum of the effects at x_1 of the smearing of all ordinates in $f(x)$; in the example shown only relatively nearby ordinates will have an effect because $g(x)$ is fairly narrow. The total output at x_1 is given by the following integration

$$h(x_1) = \int_{x'=-\infty}^{x'=+\infty} f(x')g(x_1-x')\,dx' \ . \tag{4.23}$$

The same applies for any value of x_1 and we can therefore express the output $h(x)$ as

$$h(x) = \int_{x'=-\infty}^{x'=+\infty} f(x')g(x-x')\,dx' \ . \tag{4.24}$$

This integral is described as the **convolution** of $f(x)$ and $g(x)$, both of which can be complex-valued functions. For short, it is denoted as

$$h(x) = f(x) \circledast g(x)$$

or simply $\quad h(x) = f \circledast g \ .$

Other names for convolution, though this is the most commonly used, include **folding product** (cf. the German *Faltung*), **composition product, superposition integral**. In the appropriate context $g(x)$ would be referred to as a **smoothing** (or **blurring**) **function**, *line-spread function* (LSF) or, in two dimensions, the *point-spread function* (PSF).

The example we have been considering illustrates how convolution is the distribution of one function in accordance with a law specified by another function. It involves multiplying each ordinate of a function by the whole of another function and summing the results. It should be noted that it is assumed that, as often occurs in practice, we are dealing with a **linear system** – by which it is meant that its output is a linear superposition of the outputs resulting from each individual component in the input. Also note that we have assumed 'invariance' of the function $g(x)$. We return to these matters in §5.1.

So far, this has been essentially a physical picture of what happens. To calculate the output at any particular value of x, such as x_1, it is necessary to perform the integration in Eqn (4.24). This is represented graphically in Fig. 4.05(e) where x' is the dummy variable of the integration. $f(x')$ is multiplied by $g(x_1 - x')$, and the area (shaded) under the product curve is the value, $h(x_1)$, of the output at x_1. Note that before multiplication and integration, $g(x)$ is both displaced and reflected (cf. the term 'folding product' as an alternative to convolution). A consequence of this reversal is that convolution is commutative, i.e.

$$f \circledast g = g \circledast f \ .$$

It is also associative

$$f \circledast (g \circledast h) = (f \circledast g) \circledast h$$

and distributive over addition

$$f \circledast (g + h) = (f \circledast g) + (f \circledast h) \ .$$

4.4.2 The grating as a convolution

The set of δ-functions specifying the grating lattice can be represented by $f(x)$ in Fig. 4.06, and the single-aperture function by $g(x)$, though they could equally well be chosen the other way round. We now have the result then, that the aperture function of the whole grating is the convolution, $h(x)$, of the one with the other.

This example, like those in the previous section, illustrates how convolution may be visualized as the nett effect of distributing one function in accordance with a law specified by another function. In the previous example (Fig. 4.05) the distribution causes overlapping of ordinates to some extent, whereas with the grating example it does not. A quite different example of convolution occurs in Chapter 5 in connection with image formation.

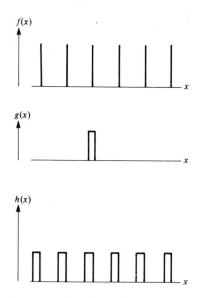

Fig. 4.06 − A grating-aperture function, $h(x)$, as a convolution of a δ-function array, $f(x)$, with the single-aperture function, $g(x)$.

4.5 THE CONVOLUTION THEOREM AND DIFFRACTION

We can now bring together some of the results of the two previous sections and obtain a result that is of considerable importance.

In §4.4.2 it was seen that the aperture function of a whole grating can be described as the convolution of one single aperture function with the array of δ-functions that specifies the distribution of that aperture in the grating.

In §4.3 we saw that the grating diffraction pattern is given by the product of the Fourier transform of the single-aperture function and the transform of the array of δ-functions defining the grating.

Since the grating diffraction pattern is the Fourier transform of its overall aperture function, we can therefore say that the Fourier transform of the convolution of the single aperture function with the δ-function array is equal to the product of the individual transforms. This is an example of the **convolution theorem** which states that the Fourier transform of the convolution of two functions is equal to the product of their individual transforms.

In the present context this leads to the very important result that convolution in object space (real space) corresponds to multiplication in diffraction space (i.e. Fourier, or reciprocal, space). This is not only of interpretative value but it also provides a powerful tool in the computational side of image processing (§5.5).

As the convolution theorem has been arrived at here in a rather special way, the reader may like to have a direct derivation.

We have used $h(x)$ to denote the convolution of two functions $f(x)$ and $g(x)$. Their individual Fourier transforms are conventionally written as $H(u)$, $F(u)$ and $G(u)$ respectively, where x and u are the usual conjugates.

The convolution theorem can therefore be expressed by saying that if

$$h(x) = f(x) \circledast g(x)$$
$$\text{then} \quad H(u) = F(u)\,G(u) \, . \tag{4.25}$$

To show this, we start by using the definition of a Fourier transform to write

$$H(u) = \int_{-\infty}^{+\infty} h(x)\, e^{\,2\pi i u x}\, \mathrm{d}x \, . \tag{4.26}$$

Substituting for $h(x)$ as the convolution of $f(x)$ and $g(x)$, using Eqn (4.24), gives

$$H(u) = \int e^{\,2\pi i u x} \left[\int f(x')\, g(x - x')\, \mathrm{d}x' \right] \mathrm{d}x \, . \tag{4.27}$$

Changing the order of integration this becomes

$$H(u) = \int f(x') \left[\int e^{\,2\pi i u x}\, g(x - x')\, \mathrm{d}x \right] \mathrm{d}x' \, . \tag{4.28}$$

In the inner integral, in which x' is constant, we make the change of variable $x - x' = X$, then $dx = dX$ and we have

$$H(u) = \int f(x') \left[\int e^{2\pi i u(X+x')} g(X) dX \right] dx' . \qquad (4.29)$$

The factor $e^{2\pi i u x'}$ is a constant so far as the inner integral is concerned, and as the rest of this inner integral does not contain x' it is a constant in the outer integration with respect to x'. Therefore the whole breaks up into the product of two separate integrals, giving

$$H(u) = \int_{-\infty}^{+\infty} f(x') e^{2\pi i u x'} dx' \int_{-\infty}^{+\infty} g(X) e^{2\pi i u X} dX \qquad (4.30)$$

i.e. $\qquad H(u) = T[f(x)] \times T[g(x)] \qquad (4.31)$

or $\qquad H(u) = F(u) G(u) \qquad (4.32)$

where T denotes Fourier transformation.

4.6 FOURIER TRANSFORMS AND LIGHT WAVES

The comments made in §1.2 about the wave nature of light and the discontinuous nature of its production can now be consolidated. To do this, the Fourier transform is used in just the same way as in studying the relationship between an aperture function and the distribution of its constituent spatial frequencies. Instead of what can be regarded as a 'space pulse' we now have a pulse in time associated with a distribution (i.e. spectrum) of temporal frequencies. The conjugates x and u are replaced by time, t, and frequency, ν. The Fourier transform equations can be written down directly by changing the symbols in Eqns (4.08) and (4.09) as follows

$$f(t) = \int_{-\infty}^{+\infty} F(\nu) e^{2\pi i \nu t} d\nu \qquad (4.33)$$

$$F(\nu) = \int_{-\infty}^{+\infty} f(t) e^{-2\pi i \nu t} dt . \qquad (4.34)$$

To illustrate the analogy between this new scenario and diffraction, Fig. 4.07(a) shows a rectangle function and its transform, now labelled according to the new variables. However, as we already know, a major component of this rectangle function is a non-oscillatory (i.e. zero frequency) constant amplitude — due to the function being entirely positive. An example more appropriate to a consideration of light waves is the transform pair shown in Fig. 4.07(b).

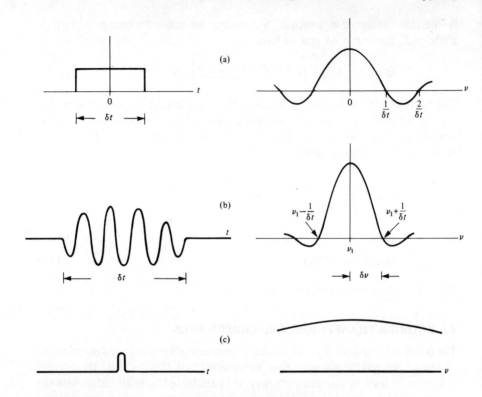

Fig. 4.07 − Time-frequency Fourier transforms.

This shows that if a pure sinusoidal wave of frequency ν_1 is truncated to form a wavetrain of finite duration and length, then it is composed of an amplitude-frequency distribution spread about ν_1 such that the summation gives a **wave group** (or **wave packet**) which represents the profile within the wavetrain but becomes out of step to give a zero nett amplitude on either side of it. When the wavetrain is long the frequency spread is narrow, and vice versa: there is the same reciprocity as with the spatial Fourier transform pair. Strictly monochromatic light would involve wavetrains of infinite length and as this is clearly not physically realizable, especially with light emitted discontinuously as **photons** by atoms, all spectral lines have a finite width. If, in Fig. 4.07(b), the 'width' of the frequency distribution is taken as essentially within $\nu_1 \pm \delta\nu$ we have

$$\delta\nu \cdot \delta t \approx 1 \ . \tag{4.35}$$

This measure of width, the **bandwidth**, is used because most of the energy is associated with the central peak, as can be seen if the amplitude is squared to give the energy distribution (the **power spectrum**).

Since with **thermal** (i.e. non-laser) light sources the wavetrains emitted are not of identical duration, and the frequencies are modified by thermal motions, field effects, etc., it is necessary to think of an average duration time, Δt, and a spectral distribution somewhat different from that considered above. However, the reciprocity between pulse duration and frequency spread is fundamental and if the frequency bandwidth is $\Delta \nu$ we have the so-called **Bandwidth Theorem**

$$\Delta \nu \cdot \Delta t \approx 1 \ . \tag{4.36}$$

Δt in this equation is the **coherence time** of the light. The corresponding **coherence length**, l_c, is given by

$$l_c = c \cdot \Delta t \approx c/\Delta \nu \ . \tag{4.37}$$

Using $c = \lambda \nu$ this becomes

$$l_c \approx \lambda^2/\Delta \lambda \ . \tag{4.38}$$

Light sources such as discharge tubes give spectral linewidths of the order of 0.1 nm so that, taking $\lambda \approx 550$ nm, $l_c \approx 5500$ wavelengths. Although this is a large number of wavelengths it is physcially a short length (about 3 mm) compared with the dimensions and displacements used in interferometry. And in Chapter 6 we indeed see how Fourier methods in spectroscopy are based on measurements of temporal coherence by interferometry.

In contrast to the above example, the linewidth of **laser light** is so narrow that l_c can be tens of kilometres, whereas the **'white light'** from a tungsten filament lamp has such a wide frequency range (Fig. 4.07(c)) that it consists of pulses of extremely short duration with no one frequency predominating (— used by some authors as a definition of a **pulse**).

4.7 CORRELATION

Correlation is one of the most ubiquitous of all signal (optical and other) and data treatment tools in use today; and we see examples of its applications in Chapters 5 and 6. In its various manifestations it is essentially a method for assessing and specifying mutual relationships where these take the form of similarities or coincidences. An example is the 'correlation' between the phases of the Moon with the tides on the Earth: plotted as a function of time the two periodicities 'correlate'.

The process of testing for a correlation is basically one of comparing patterns; but it is not as straightforward as it may seem at first. An example will reveal the problem that arises. In Fig. 4.08(a) a pattern in the form of a 'motif' is located in the x,y plane. Let it be described by a function $f(x,y)$, where f is some chosen property of the motif, e.g. its colour, composition, reflectance, transmittance, etc.; outside the motif f is zero. This function is no use, however, for describing the motif independently of its position in the x,y plane. Yet in many instances

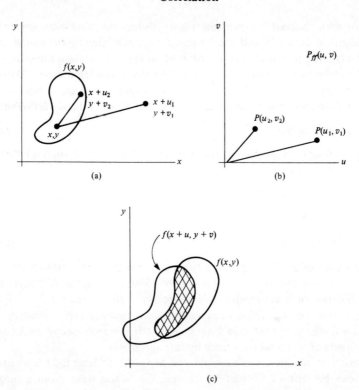

Fig. 4.08 – Autocorrelation.

a way of specifying a function is needed that is 'invariant under translation'; one that has coordinates that are internal to the function but specify it as well as does $f(x,y)$. Consider a function $P_{ff}(u,v)$ that is the product of the value of f at x,y multiplied by its value at $x + u$, $y + v$, the product being summed (integrated) for all x,y. This can be expressed as

$$P_{ff}(u,v) = \int\!\!\int_{-\infty}^{+\infty} f(x,y) f(x + u, y + v) \, dx \, dy \quad . \tag{4.39}$$

As illustrated in the figure, this function is a vectorial description of the motif. Since f is zero for all points outside the motif, in (a) the product of the values of f at pairs of points in the x,y plane separated by u_1, v_1 and summed for all x,y is zero. The value of P at u_1, v_1 in (b) is therefore zero. On the other hand the value of $P(u_2, v_2)$ is clearly non-zero.

In this way P maps out a pattern that is a description of f independent of its position in the x,y plane. As this procedure amounts to a comparison of a function with itself, Eqn (4.39) is referred to as the **autocorrelation function**. $P(0,0)$ is obviously a strong peak in the P map and it can be proved that it has its maximum value there.

It is important to note that Eqn (4.39) also gives us another way of visualizing the calculation. It shows that the value of P for any chosen u,v is obtained by shifting the function f with respect to itself by $-u,-v$ and determining the area of overlap (Fig. 4.08(c)) (cf. the comment above about $P(0,0)$ being the maximum value of P).

To compare two different motifs, $f(x)$ and $g(x)$, we can write a **cross-correlation function**

$$P_{12}(u,v) = \int\int_{-\infty}^{+\infty} f(x,y) g(x+u, y+v) \, dx \, dy \tag{4.40}$$

where P now refers to products between f and g.

Both correlation functions can be expressed in terms suitable for one, two or three dimensions and for time. To explore them in a little more detail we will revert to using one dimension:

$$P_{12}(u) = \int_{-\infty}^{+\infty} f(x) g(x+u) \, dx \tag{4.41}$$

and similarly for P_{11}.

In order that u can describe a shift in the positive x direction the above equation can be rewritten as follows:

$$P_{12}(u) = \int_{-\infty}^{+\infty} f(x) g(x-u) \, dx \; . \tag{4.42}$$

A simple 1-dimensional example of autocorrelation is shown in Fig. 4.09(a) and (c). ((b) and (d) refer to §4.7.1.) The reader should draw a few sketches to show that when a second rectangle function, identical to the one in (a), is shifted by various amounts along the x-axis, the area under the product for each shift builds up the result shown in (c).

The definition of correlation is often expressed in a slightly different form from the above, as follows:

$$c(x) = f(x) \odot g(x) = \int_{-\infty}^{+\infty} f(x') g(x'-x) \, dx' \tag{4.43}$$

and similarly for autocorrelation. In this formulation x is the shift and x' is the dummy variable for integration.

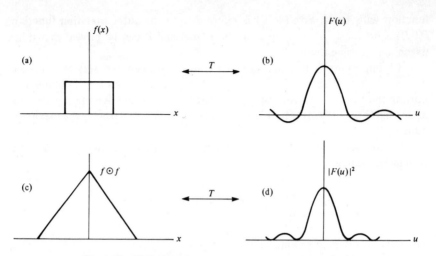

Fig. 4.09 — Example of:
autocorrelation — (a) (c)
Wiener–Khinchin theorem — (c) (d).
T denotes Fourier transformation.

Correlation is therefore like convolution (cf. Eqn (4.24)) but differs in that g is shifted but not reversed (folded). The symbol $\odot$ is used fairly widely to distinguish between correlation and convolution, though unfortunately there are no universally adopted symbols for these operations.

When the functions are complex it is customary to define the **complex cross-correlation function** as

$$\gamma_{fg}(x) = f(x) \odot g^*(x) = \int_{x'=-\infty}^{x'=+\infty} f(x') g^*(x'-x) \, dx' \qquad (4.44)$$

and the **complex autocorrelation function** as

$$\gamma_{ff}(x) = f(x) \odot f^*(x) = \int_{x'=-\infty}^{x'=+\infty} f(x') f^*(x'-x) \, dx' \ . \qquad (4.45)$$

(Note that if the asterisk is associated with the other function the conjugate result is obtained.)

For $f(x)$ and $g(x)$ complex, the difference between convolution and correlation referred to above can be summarized as follows:

$$f(x) \odot g^*(x) = f(x) \circledast g^*(-x) \ . \qquad (4.46)$$

It follows that correlation is not commutative i.e.

$$f(x) \odot g^*(x) \neq g^*(x) \odot f(x) \ . \qquad (4.47)$$

For some purposes it is convenient to normalize correlation functions and this is done by dividing the expressions in Eqns (4.44) and (4.45) by the central value of the correlation, i.e. by the value for $x = 0$ (zero shift). This gives

$$\text{Normalized } \gamma_{fg}(x) = \frac{\int f(x') g^*(x'-x) \, dx'}{\int f(x') g^*(x') \, dx'} \tag{4.48}$$

$$\text{Normalized } \gamma_{ff}(x) = \frac{\int f(x') f^*(x'-x) \, dx'}{\int f(x') f^*(x') \, dx'} \tag{4.49}$$

$$= \frac{\int f(x') f^*(x'-x) \, dx'}{\int |f(x)|^2 \, dx'} \tag{4.50}$$

where the integrations are from $-\infty$ to $+\infty$.

4.7.1 Autocorrelation theorem (Wiener–Khinchin theorem)

This theorem states that the Fourier transform of the autocorrelation of a function $f(x)$ is the squared modulus of its transform.

Taking the general case, if the transform of a complex function is $F(u)$ the theorem means that the transform of $|F(u)|^2$ is the complex autocorrelation of $f(x)$.

Using T to denote Fourier transformation, we have

$$T[|F(u)|^2] = \int_{-\infty}^{+\infty} |F(u)|^2 e^{2\pi i u x} \, du$$

$$= \int_{-\infty}^{+\infty} (F(u) F^*(u)) e^{2\pi i u x} \, du$$

$$= f(x) \circledast f^*(-x)^\dagger \tag{4.51}$$

---

†The negative sign in $f^*(-x)$ arises from Fourier transformation of a conjugate:

$$\int_{-\infty}^{+\infty} F^*(u) e^{2\pi i u x} \, du = \left[\int_{-\infty}^{+\infty} F(u) e^{-2\pi i u x} \, du \right]^*$$

i.e. $\quad T[F^*(u)] = f^*(-x)$.

whence, using Eqn (4.46),

$$T[|F(u)|^2] = f(x) \odot f^*(x) \tag{4.52}$$

i.e. $\qquad F(u)\, F^*(u) \xrightarrow{T} f(x) \odot f^*(x) \tag{4.53}$

This is illustrated in Fig. 4.09 for the real function $f(x)$.
Changing to time and frequency gives

$$T[|F(\nu)|^2] = f(t) \odot f^*(t) \tag{4.54}$$

where $|F(\nu)|^2$ is described as the **power spectrum** in terms of amplitude squared, as a function of (*temporal*) *frequency*. By analogy, $|F(u)|^2$ in Eqn (4.52) is referred to as a **power spectrum** in terms of *spatial frequency*. Applications of both await us in the chapters ahead.

5

Optical imaging and processing

5.1 INTRODUCTION

The major Fourier aspects of optical image formation and processing under coherent and incoherent illumination are outlined in this chapter. They involve Fourier transforms, alone or in association with convolution and correlation. Attention must be drawn without delay, however, to the fact that their importance is not restricted to the handling of data that are optical in origin. There are numerous instances today where optical processing is used even when the initial data are not optical. The basic reason for this is the way in which the same mathematical operations that apply to most optical systems also often apply to electrical communications systems. The optical analogue is attractive because it has the advantage of 2-dimensional display and parallel processing; and it is being made increasingly practicable by the development of electrical-to-optical interface devices, in conjunction with the computer. When, for some reason, practical optical techniques are not used, the computer can be used alone to enable the same fundamental principles to be employed for 'digital imaging' and processing.

Our purpose in this chapter is, however, solely to deal with the essentials of how Fourier transforms, convolution and correlation, already introduced in the optical context in previous chapters, form much of the basic tool-kit for optical imaging and processing. For detailed information on particular facets and the innumerable applications in the field of communications, the specialist books should then be more accessible.

Before embarking on these topics in detail it will be helpful to have a descriptive preview of the sections (5.2 and 5.3) on optical image formation itself. Optical processing, as opposed to imaging, is concerned with intervention in the process in various ways and for various purposes, and that is the subject of §5.5.

There are two models of optical imaging that are useful today, each with applications to which it is particularly well suited. Both were touched on in Chapter 1 and have been mentioned to varying extents in the intervening chapters.

The *first model* places emphasis on the overall diffraction (scattering) of light from an object when the conditions are at least partially coherent, and the way in which that light is recombined to form an image. The Fourier aspects of the first part of this are already familiar from Chapters 3 and 4. In §5.3 we consider this again, particularly in relation to the second stage of image formation. This model was originally formulated essentially qualitatively by E. Abbe in 1873 to deal with his experimental observations on the imaging of periodic objects under the microscope. In today's terminology he found that with the illumination methods normally used in optical microscopy, image formation is not the completely incoherent process sometimes assumed; in fact it can be nearly coherent with some modern systems.

For a coherently illuminated object the quantitative identification of diffraction orders with the terms of the Fourier series analysis describing its structure, and the way an imaging lens recombines them as a Fourier synthesis to form an image, were reported in some detail by A. B. Porter in 1906. (It does not seem to be widely appreciated that a significant step in that direction had already been taken by Lord Rayleigh in 1874, and certainly in his paper of 1896.)

The extension of the Abbe–Porter interpretation to include the analogous role of Fourier transforms in the coherent imaging of non-periodic objects was made mainly by X-ray crystallographers by the early 1940s.

The value of the approach offered by this model is its sensitivity to the way in which spatial frequencies in an object 'structure' (periodic or non-periodic) are conveyed by diffracted wavefronts and reconstituted to form an image. Using coherent illumination it leads to methods for intervention in the diffraction plane (the spatial frequency plane) so that the formation of the image can be controlled there by 'filtering'. This is one aspect of optical processing; others are mentioned in §5.5.

The *second model* of image formation, which we consider in §5.2, is applicable to both coherent and incoherent illumination conditions. Here again Rayleigh made an important contribution (1896), this time with influences from the earlier work of Airy and Helmholtz. The model visualizes an image as the combination of the Airy patterns (or more complicated patterns if aberrations are present) that an optical system would separately produce with the light leaving every individual point in the object. If the illumination is incoherent the Airy intensity patterns due to all the object points are simply additive. If it is coherent there is interference and mathematically one deals with the combination of the complex-amplitude Airy patterns. Rayleigh dealt with both extremes. By considering an object in the form of a row of equally spaced point sources he used a Fourier series to represent the summation of the regularly spaced Airy intensity patterns forming the image according to this model. For coherent conditions he showed that the terms of the series correspond to the diffraction orders of the Abbe model referred to above.

An important development of this **convolution model** of image formation is

[§5.1]   **Introduction**   85

due to Duffieux who, in his book *L'Intégrale de Fourier et ses applications a l'optique,* acknowledged his indebtedness to 'Michelson et surtout Lord Rayleigh' as the founders of the use of Fourier methods in physical optics. We shall read a great deal more about Michelson in Chapter 6. And of Rayleigh, that most versatile scientist who made major contributions in nearly every branch of physics, we can note that he gained his Nobel Prize in 1904 for physics, for his investigations on the densities of the more important gases and for his discovery of argon.

Duffieux appears to have been the first formally to express the image of a continuous object distribution as a convolution (he used the term *'faltung'*) of that distribution with what we call the point-spread function, or impulse response of the system (the Airy pattern in the simplest case), using complex amplitudes or intensities as appropriate for the illumination conditions. Application of the convolution theorem then showed him that the spatial frequency spectrum of the image is the product of the frequency spectrum of the object distribution and that of the response of the system. The optical system could therefore be considered as 'transferring', to different extents, the spatial frequency components of the object to the image plane. This work gave us one of the most important concepts in this subject today, namely that of an optical imaging system as having a **transfer function** (Duffieux's *'facteur de transmission'*) for each frequency entering it. When it is applicable — and we shall come to that in a moment — it provides an invaluable approach to matters concerned with the design and analysis of optical systems. Its virtues may be seen as complementary to those of the Abbe model.

Another influence in establishing the convolution model of imaging has come from the work of Schade (1948), Elias *et al.* (1952) (see also Elias, 1953) and others who have pointed to the link that exists between these aspects of optics and the ideas and methods that had been developed in connection with the analysis of electrical networks and other communications problems of a **linear systems** type. A short digression here into this other area is justified by the valuable cross-fertilization of ideas that has resulted from the recognition of the way in which the two share the same fundamental principles.

In the general sense a 'system' may be defined as any device that 'maps a set of input functions on to a set of output functions'. And as noted in §4.4.1, in slightly different terms, a system is 'linear' if its response to a number of input signals is the sum of the responses obtained when the signals are applied separately. In the optical context a system is linear if, as is commonly the case, spatial frequencies are transmitted through the system with alteration only of their amplitude and phase, not their frequencies.

In the design and analysis of linear electrical networks one method had been to consider an output in just the same way as we have described above for optical image formation, viz. a convolution of the input (expressed as a series of impulses of changing magnitude) with the unit impulse response of the system.

The integrations involved when the effects of different filters were being investigated, however, became very complicated. Even more difficult was the deconvolution needed in the design of filters required to produce specified outputs from predetermined inputs. It was the application of the convolution theorem that produced the much needed simplification in many instances. From the theorem it follows that the temporal frequency spectrum of the output from a linear electrical network system is simply the product of the frequency spectrum of the input and the frequency spectrum of the unit impulse response of the system (its transfer function). Integration in the time domain is replaced by the simpler operation of multiplication in the frequency domain. Furthermore, the overall frequency response of several filters in tandem is simply the product of their individual transfer functions. Small wonder that it was said that if network theory had been limited to the time domain approach, it would never have advanced.

However, the straightforward use of the convolution theorem as described above, whether in communication systems or in optical imaging systems, necessitates an additional requirement, namely that of **invariance** (or 'stationarity'). Strictly speaking, this means that in an electrical network, for example, the response to a unit-impulse must be independent of the time at which that input is applied — i.e. it is a *time-invariant system*. Likewise, in an optical imaging system the image of a point object — the point-spread function — must be the same over the entire field; it must be a **space-invariant** system (cf. §4.4.1). Early in the following section we will discuss the implications of this in optical image processing. (The way in which a system that is not *'invariant linear'* is dealt with, and indeed the problems of non-linear systems in general, are outside the scope of this book.)

Electrical engineering has had other influences on optics, notably arising from the work initiated and developed by Wiener (1949) and Shannon (1949) on statistical network theory and information theory. This has led, for example, to the use of correlation functions in the real-space processing of optical images, some aspects of which are introduced in §5.5.

5.2 INCOHERENT OPTICAL IMAGING

The description in the previous section of image formation as a convolution can usually be applied to the normal use of a telescope or camera etc., as a convolution of intensities. For most photographic, television, and other systems it has been found that there is good agreement between experimental measurement and calculation based on the assumption that illumination under normal conditions is essentially incoherent (Barnes, 1971). Happily, there are a number of advantages in this, including the opportunity of using TV pictures, LED displays, etc., as inputs to processing systems.

When optical aberrations are present in the system, however, the unit impulse response of the system, which is here the point-spread function of the system (§2.3), may well be different for different points in the object field. Such variations could, as we have seen, invalidate the use of the convolution theorem. Fortunately, when a system is well corrected the residual effects of aberrations are reasonably constant over the area within which the image of any point in the object field is of significant intensity. The system is then said to be **isoplanatic** (analogous to it being 'achromatic' when the positions of image points are independent of the wavelength of the light in the wavelength range concerned). Under those circumstances the system is for practical purposes space-invariant, and the convolution theorem can be used in a straightforward way.

As indicated in Fig. 5.01, though purely symbolically in 1-dimension, the application of the convolution theorem gives the frequency spectrum of the image intensity distribution as the product of the frequency spectrum of the object intensity distribution and the Fourier transform of the PSF. The transform of the PSF is the optical transfer function (OTF) of the system.

This means that every sinusoidal component in the object distribution is 'transferred' to the image without alteration of its frequency: only its intensity and phase are subject to alteration by the system. Consider the intensities first. Figure 5.02(a) represents an object intensity distribution containing a single sinusoidal component of some specified spatial frequency. The 'strength' of this component is expressed in terms of the modulation M, given by

$$M = B/A \tag{5.01}$$

(cf. definitions in §1.1 and Eqn (1.05)).

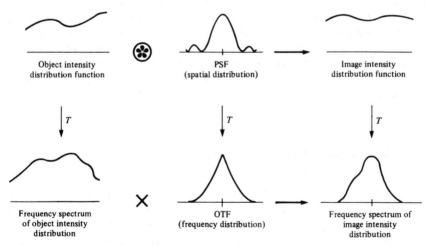

Fig. 5.01 — Spread and transfer functions. Schematic diagram of functional relationships for incoherent illumination.
T, Fourier transform; PSF, point-spread function: OTF, optical transfer function; ⊛ convolution; × multiplication.

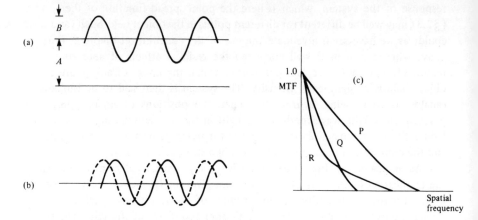

Fig. 5.02 – Transfer functions (incoherent illumination).
(a) Modulation of an individual frequency component is given by B/A;
(b) associated phase shift.
(c) examples of modulation transfer functions.

The ratio of the modulation in the image to that in the object is the modulation transfer *factor* for that frequency: effectively an index of the efficiency of the imaging system at that frequency, it is usually less than unity (except for zero frequency). Expressed as a function of frequency over the appropriate range it comprises the **modulation transfer function** (MTF) of the system.

Any frequency-dependent phase changes introduced by the system would be manifest in the image as lateral shifts of the sinusoidal components comprising the image, as shown in Fig. 5.02(b). Complementary to the MTF, there is therefore a **phase transfer function** (PTF) though this is often negligible.

The two functions comprise the optical transfer function (OTF) (or **frequency response**) of the system as follows

$$(\text{OTF}) = (\text{MTF}) \, e^{i(\text{PTF})} \quad . \tag{5.02}$$

5.2.1 Determination of transfer functions

(i) Calculation

Several methods are available for calculating the OTF of an optical system from its design data. For the contribution due to aberrations one method is to compute the passage of a large number of rays through the system from a single object point. By spacing the rays uniformly across the aperture of the lens, the density distribution of resulting points in the image plane gives the intensity distribution that comprises the point-spread function. Fourier transformation then gives the **geometrical** OTF of the system. If a system is free from aberrations the geometrical OTF would be unity for all frequencies; every object point would be imaged as a point. Correction for diffraction is made by multiplying this geometrical transfer function by the transfer function for the equivalent diffraction limited system, i.e. the system free from all imperfections.

Other methods involve the detailed computation of the aperture function, including the effects of aberrations. This is a complex amplitude distribution over the aperture and we will denote it by $f(x)$ as for aperture functions in previous chapters. Its Fourier transform, $F(u)$, is the complex amplitude diffraction pattern of the image of a point source. The squared modulus of this is the PSF, and the Fourier transform of that is the OTF. This is illustrated symbolically in 1-dimension in Fig. 4.09 for the familiar example of $f(x)$ as a simple rectangle function. The calculation route just described is (a) → (b) → (d) → (c).

Alternatively, writing

$$\text{OTF} = T\{|F(u)|^2\}$$

where T denotes Fourier transformation, and using the autocorrelation theorem (§4.7.1), gives

$$\text{OTF} = f(x) \odot f^*(x) \tag{5.03}$$

i.e. the OTF is given directly by calculating the autocorrelation of the aperture function. In Fig. 4.09 this is route (a) → (c).

(ii) Experimental

Various methods, some quite complicated, are employed in the experimental determination of transfer functions. Only a simple method, chosen because it re-emphasizes the meaning of what is involved, need be mentioned here. In this the MTF is determined for one frequency at a time. A series of object screens is used, each bearing a sinewave pattern of different spatial frequency; they look rather like an out-of-focus shadow of a row of evenly spaced parallel knitting needles. The ratio of the modulation in the image to that in the object is determined for patterns of spatial frequencies covering the necessary frequency range. Frequency-dependent phase shifts, comprising the PTF, are given by the relative lateral positions of the image and object bands at each calibration frequency. By its nature this method gives the 'line-spread function' (LSF) rather than the point-spread function (PSF).

These transfer functions provide a much more informative appraisal of a lens system than does just a measurement of its resolution limit. This is illustrated by the MTF curves in Fig. 5.02(c). Curve P corresponds to a lens free from all aberrations, the contrast ratio decreasing with increasing frequency until it reaches zero at the resolution limit of the lens (cf. Fig. 5.01). Curves Q and R are representative of lenses with aberrations; they show that whilst R has a frequency limit superior to Q, it gives an image contrast (modulation) that is inferior at lower frequencies; choice between the two can be made according to the type of application. Though greatly superior to the older and rather misleading measurement of resolution limit as a criterion of optical performance, optical transfer functions are not the complete answer to the problem of assessing performance, especially if the eye is involved in the final imaging. The eye is a relatively poor imaging system but it is associated with complicated data-processing in the retina and brain. This makes the prediction and determination of its overall response in any particular situation very difficult.

5.3 COHERENT OPTICAL IMAGING

Ernst Abbe's experimental work in the 1870s on improving the performance of microscope objectives laid the foundation of the approach used today in considering coherent imaging. Working at the University of Jena in collaboration (later in partnership) with Carl Zeiss at the latter's optical works, Abbe found that microscope objectives made with the most careful correction of aberrations gave resolution inferior to less carefully corrected objectives of larger aperture. In experiments with periodic specimens such as insect scales and diatom skeletons he showed that the influence of a large aperture is associated with diffraction at the specimen itself (— evidence of some coherence in the illumination conditions).

He demonstrated how the diffraction maxima formed in the back focal plane of an objective contribute to the formation of the image, the higher orders (or higher spatial frequencies as we now regard them) controlling the fine detail in the image. Abbe had introduced wave theory ('wave optics') into instrumental optics, previously the exclusive province of geometrical (or 'ray') optics.

The interpretation of Abbe's work in terms of Fourier series was described in an important paper by A. B. Porter in 1906. And in an elegant series of experimental demonstrations he also showed their physical reality. For example, he showed the effects on the image of a periodic object when different combinations of diffraction orders leaving the object are prevented from contributing to the image.

Subsequently extended by the application of Fourier transform methods, to deal with the imaging of non-periodic objects, the Abbe approach has been put to use in a number of outstandingly important ways. As already mentioned, these mainly depend on the spatial frequency aspects of Fraunhofer diffraction and the accessibility of the diffraction pattern, mathematically and experimentally, if coherent conditions are used.

5.3.1 Periodic objects

Consider the 1-dimensional multiple-slit transmission grating of earlier chapters, acting as an object being imaged by a lens (Fig. 5.03). Wavefronts constituting the various diffraction maxima are brought to separate foci in the back focal plane of the lens, and it is the light that passes through these foci in the diffraction

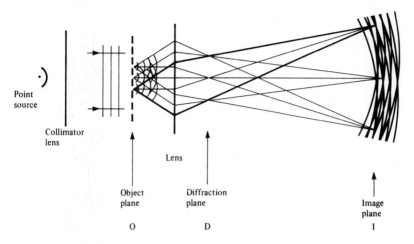

Fig. 5.03 — Abbe's treatment of image formation in the microscope (after Bragg, 1929).

plane that combines in the image plane to give an optical reconstruction of the object: for the moment we will assume the lens to be perfect and of infinite aperture. Note that, as indicated by the two heavy lines in the figure, there is no conflict between this wave description and the ray-tracing method of geometrical optics.

Alone, *any* pair of the foci produces a set of sinusoidal bands in the image plane. This is reminiscent of Young's experiment (§1.1) where the pair of apertures act in the same way. In this sense, image formation can be thought of as a double process of diffraction — an approach put forward by F. Zernike in about 1935.

To see how an image is built up in this way, with the type of object grating envisaged in Fig. 5.03, we will consider the contributions to the image made by the pairs of diffracted beams comprising the various orders of diffraction.

Figure 5.04 shows, schematically, the formation of the nth order pair of diffracted beams leaving the object, and how in the image plane they interfere to produce a harmonic distribution of illumination with repeat D'_n given by

$$D'_n \sin \theta'_n = \lambda \ . \tag{5.04}$$

The condition for the formation of the nth order diffraction maxima at the object is

$$D \sin \theta_n = n\lambda \ , \tag{5.05}$$

whence
$$D'_n = \frac{D}{n}\left(\frac{\sin \theta_n}{\sin \theta'_n}\right) \ .$$

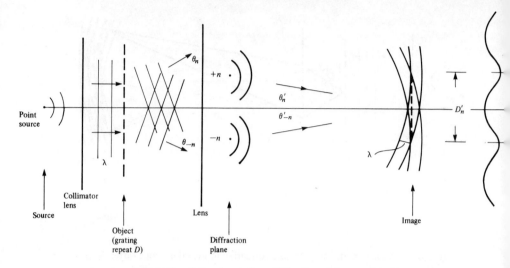

Fig. 5.04 — Image formation: diffraction and recombination.

The term in brackets is a constant related to the magnifying power of the lens. So we have

$$D'_n \propto D/n \ . \tag{5.06}$$

The first-order pair of maxima from the object interfere in the image plane to give a simple-harmonic variation of illumination that corresponds just to the basic repeat of the grating: this repeat is the bare minimum of information about the object, with no fine detail of its optical structure. Each pair of successively higher-order maxima adds a harmonic of successive shorter repeat ($\propto D/n$) to the total illumination that forms the image: the full detail of the image is built up by what is clearly recognizable as Fourier synthesis. Now in §3.4.1 it was shown that the diffraction maxima themselves comprise the Fourier analysis of the object grating, and reference was made to the diffraction plane being described as the Fourier plane. Therefore the image formation process in the example we are considering may be regarded as a double Fourier process, with the diffraction pattern as a Fourier analysis of the grating, and the image as a Fourier synthesis of the Fourier analysis. This is particularly evident when one invokes the principle of reversibility. All the diffraction orders that build up the image as a summation of harmonics return to the object grating where they recombine to form the original illumination distribution (the aperture function) at the grating.

It is important to appreciate that this double Fourier description is in complete accord with Zernike's double-diffraction interpretation; they are different ways of expressing the same thing.

For perfect imaging, an infinite Fourier synthesis would be required, necessitating not only the generation of an infinite number of diffraction orders but also their entire admission into the optical system. Both are clearly impossible. Eqn (5.05) shows how the values of D and λ impose a limit to the number of possible diffraction maxima that can be produced; and any objective of course has a finite aperture ($\sin \theta < 1$).

At the other extreme, if conditions allowed only the zero-order and first-order pair of diffraction maxima to enter the objective only the basic repeat, D, in the object grating would be resolved in the image; and then, as noted above, only as a simple sinusoidal variation of illumination. Eqn (5.05) shows that on this basis the smallest object repeat that is resolvable is given by

$$D \sin \theta = 1\lambda$$

i.e. $\qquad D = \lambda/\sin \theta \qquad (5.07)$

where $\sin \theta$ is recognizable as the **numerical aperture** of the objective (§2.3). This gives the **resolution limit** as approximately the wavelength of the illumination used.

The resolution limit can be reduced below that described above since the combination of the zero-order with only one of the first-order maxima is sufficient to form an image of the basic grating repeat. By using convergent illumination it can be arranged that these just enter opposite sides of the aperture of the objective, and the resolution limit is then $\sim \lambda/2$. This minimal need for the zero-order and one first-order maximum is the **Abbe principle.**

Exactly the same general findings apply in the case of a 2-dimensional grating, i.e. a grating in the form of a 2-dimensional lattice at each point of which is an identical aperture (§2.6). In the case of a fabricated grating each 'aperture' can, for example, be a pinhole or a group of pinholes. The diffraction pattern is then in the form of a 2-dimensional lattice of spots of illumination, the diffraction order of each spot being specified by two integers (cf. the three integers in Eqns (2.18) for a 3-dimensional grating). Recombination, the second stage in the formation of the image, is completed in exactly the same way as for the 1-dimensional example.

In §5.3.3 the experimental observation of some of these effects is described.

5.3.2 Non-periodic objects
The treatment in the previous section can be extended to deal with non-periodic objects because *discrete* orders of diffraction are not a prerequisite. A non-periodic object can be regarded as equivalent to one aperture of a grating, and we know that under those circumstances a Fourier transform instead of a Fourier series is involved. The diffraction pattern in the focal plane of the lens is one of continuous scattering, with a directional variation of amplitude and phase depending on the aperture function: it is the Fourier transform of the amplitude distribution function of the object (cf. the lens as a 'Fourier transformer' in §4.2). Recombination of this pattern in the image plane is a summation

of fringes produced by pairs of diffracted beams (at $\pm\theta'$ in Fig. 5.04) but with a continuous range of fringe spacings and orientations. Image formation can then be described as a double Fourier transformation process. This description is commonly applied to both periodic and non-periodic objects since even the former are of finite extent and it is very convenient to be able to speak of an image as being the transform of a diffraction pattern regardless of the exact nature of the object. We have already used the idea in §4.5.

5.3.3 Illustrations: optical transforms

An optical diffractometer set-up for demonstrating and using the concepts described above is shown in a simplified form in Fig. 5.05. A helium–neon laser is commonly employed, with a beam expander to provide illumination with near-perfect coherence (temporal and spatial) across the whole plane-wavefront at O where object masks are placed. The diffraction pattern (Fourier transform) produced by a mask at O is formed at D, the focal plane of the objective L_1; and the image (the double transform) of O is formed at I. In practice, to obtain diffraction patterns of reasonable size, L_1 is a long focal length lens or equivalent (e.g. a telephoto combination). A second lens, L_2 (only its position is shown (dashed line) in the figure) is then required for the formation of a real image at a reasonable distance from the object mask.

Photographic records of diffraction patterns and images are, by their nature, records only of intensities (except in holography). In the following examples the illustrations are therefore described as **optical transforms** rather than Fourier transforms. They are largely drawn from the definitive collection of books on optical transforms by H. Lipson and his colleagues (see the Bibliography).

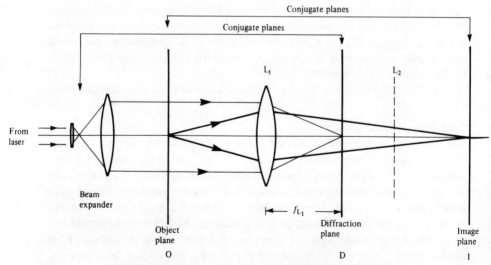

Fig. 5.05 – Basic arrangement for observing optical transforms.

(i) Diffraction by object masks

The sequence in Fig. 5.06 illustrates some of the main points that have been mentioned concerning the role of diffraction in the first stage of image formation in coherent light. Each example shows an object mask and the optical transform it gives.

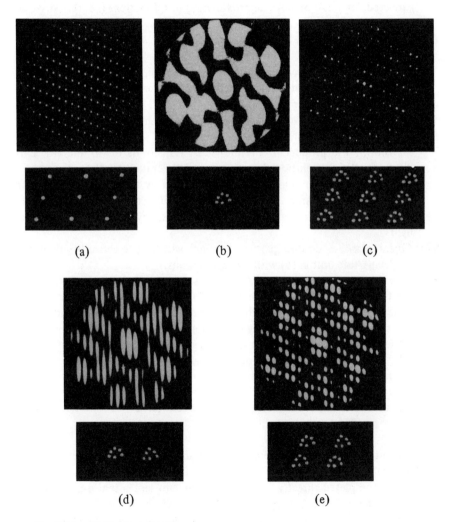

Fig. 5.06 – Optical transforms.
        Each transform is accompanied beneath by its mask, except in (a) and (c) for which the masks contained many more repeats than are shown here.
(Adapted with permission, from Taylor and Lipson (1964), *Optical transforms*. Bell, London.)

(a) *Object mask:* a 2-dimensional 'point' lattice of pinholes. It can be regarded as a lattice array of δ-functions. The mask contained very many more repeats than are shown here.

Transform (almost a 2-dimensional Fourier series because the object contains many repeats): a reciprocal lattice — its geometry and dimensions reciprocally related to those of the object lattice. Intensities are roughly constant, as expected for small pinholes approximating to δ-functions (i.e. the Airy disc is large).

(b) *Object mask:* a single aperture-unit consisting of a non-repeating, irregular group of holes.

Transform: a non-repeating pattern in which the illumination varies with direction in a way determined by the arrangement and sizes of the holes comprising the object mask.

(c) *Object mask:* the convolution of (a) and (b). As in (a), the mask contained very many more repeats then are shown here.

Transform: the product of the individual transforms of (a) and (b). This illustrates the convolution theorem. It also illustrates how the pattern given by a grating is a sampling of the pattern given by the unit comprising the grating (i.e. its transform). The sampling is in directions determined by the dimensions of the grating lattice (and, of course, by the wavelength of the light).

(d) *Object mask:* a double-aperture 'grating' which is a convolution of the single aperture unit in (b) with two δ-functions.

Transform: again the product of the two separate transforms. In this instance the transform of one unit is multiplied by that of a double-aperture grating. Note that the $\cos^2$ fringes of the latter are perpendicular to the line joining the two 'apertures' (cf. Young's experiment): their spacing is reciprocally related to the separation of the apertures.

(e) *Object mask:* a convolution of the single aperture unit in (b) with four δ-functions that define the grating lattice of (a).

Transform: the transform of the aperture unit is now multiplied by two sets of $\cos^2$ fringes that define the reciprocal lattice of (a). The sampling aspect is again demonstrated.

Elegant examples of the use of optical transforms are to be found in X-ray crystallography where, as noted in Chapter 2, the formation of images of atoms cannot be completed directly because there are no lenses that can be used to recombine the diffracted X-rays. Stopped short at the diffraction stage, only intensities are recorded and Fourier summations cannot be performed numerically or experimentally, because of the lack of phase data. In the formative years of the subject W. L. Bragg played a key role in pioneering and developing the use of Fourier and optical principles in understanding and dealing with this and other problems in X-ray crystallography. Whilst the development of the digital

[§5.3] Coherent Optical Imaging 97

computer has led to computational methods for overcoming the 'phase problem' Bragg's work made an important contribution to the wider field of optical processing. For the main literature on the development and applications of optical methods to X-ray diffraction the reader should consult the books referred to at the beginning of this section.

An example of the analogy between X-ray diffraction by a crystal and the first stage of optical image formation of a grating object is shown in Fig. 5.07. (a) is part of an optical mask representing a 2-dimensional view, in a particular direction, of the crystal structure of phthalocyanine; (b) is the optical diffraction pattern it gives (Bragg, 1944). Not only does (b) agree with the geometry of the X-ray data but the intensities also agree with the experimentally observed X-ray intensities — indicated by the numbers alongside the spots in the figure. Historically, this was a method for determining an unknown crystal structure by making 'trial' masks based on chemical and other considerations. The method was made simpler in later developments of the technique (referred to at the end of (ii)) when it was shown that the basic unit and only three repeats are sufficient as a mask since these define the lattice on which the 2-dimensional projection of a crystal is based (Hanson and Lipson, 1952). This is illustrated in Fig. 5.06. In (c) a large number of repeats (many more than shown in the figure) of the basic unit were used as the mask, whereas in (e) only the four units defining the lattice repeats were used. Comparison of the optical transforms shows that

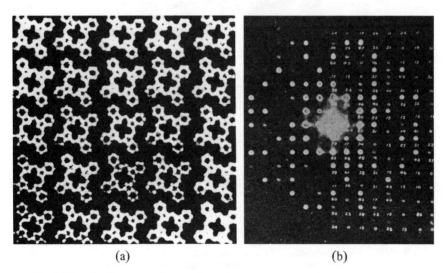

(a) (b)

Fig. 5.07 — The first stage of image formation.
(a) Mask representing a 2-dimensional projection of the crystal structure of phthalocyanine; (b) optical diffraction pattern given by (a).
(Reproduced by permission from *Nature,* **154,** 69. Copyright © 1944 Macmillan Journals Limited.)

(e) is quite adequate, compared with (c), in giving a sampling of the transform of the single unit (b) at the reciprocal lattice points (the transform) of the basic lattice (a).

(ii) Recombination

Figure 5.08 is another example from X-ray crystallography, this time of the second stage of image formation, the recombination stage. (a) is a diagrammatic 'view' down one axis of the crystal structure of the mineral diopside, $CaMg(SiO_3)_2$, which W. L. Bragg used on several occasions to illustrate the optical principles underlying the 'X-ray analysis' of crystal structure (Bragg, 1929, 1939, 1942, 1944). The atoms are represented by black discs, the larger ones representing Ca and Mg atoms which are superimposed in this view, the smaller ones representing the Si and O atoms (which form chains running through the structure). In (b) we have a mask containing small holes to represent the experimentally observed X-ray diffraction data; the areas of the holes are made proportional to the square-roots of the X-ray intensities. (c) is the image formed as the optical transform of (b) and it is a very realistic match to (a). The recombination has succeeded because

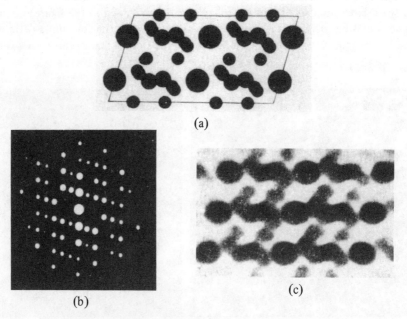

Fig. 5.08 – The second stage of image formation.
    (a) Representation of a 2-dimensional projection of the crystal structure of diopside, $CaMg(SiO_3)_2$; (b) optical mask representing the corresponding X-ray diffraction pattern; (c) optically reconstructed image using (b).
    (Reproduced by permission from *Nature*, **143**, 678; **149**, 470. Copyright © 1939 and 1942 Macmillan Journals Limited.)

nearly all the X-ray diffracted beams have the same phase, being largely determined by the strongly scattering Ca and Mg atoms which are at centres of symmetry. In examples where different phases are involved this can be allowed for by placing a piece of mica in front of each hole and rotating the pieces individually to produce the appropriate phase shifts. (Contrary to first impression this method does not require the use of polarized light in the optical reconstruction of the image.)

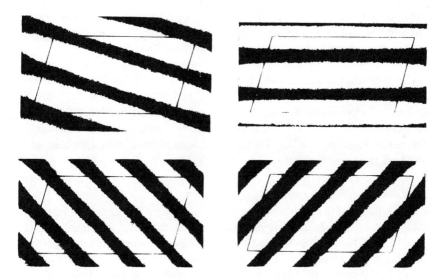

Fig. 5.09 — Sinusoidal bands of light and shade formed in the image plane by pairs of diffraction maxima. (Reproduced by permission, from Bragg (1933), *The crystalline state.* Bell, London.)

That such images can be regarded as composed of sinusoidal harmonics produced by pairs of diffraction maxima, as described in §5.3.1, had earlier been demonstrated by Bragg. Figure 5.09 shows sets of sinusoidal bands like those produced separately by pairs of different diffraction maxima, their phases manifest by the positions of their 'crests' and 'troughs' with respect to the origins at the corners of the unit cell. Bragg exposed a sheet of photographic paper successively to artificially made patterns like these to represent the diffraction effects produced by the mask in Fig. 5.08(b). Keeping exposures down to a level such that the general level of photographic blackening did not build up too much he obtained a picture like that in Fig. 5.08(c).

• • • •

The use of X-rays in the first stage of image formation followed by the use of visible light and a lens to complete the second stage became known (Buerger, 1950) as **2-wavelength microscopy,** although Bragg preferred 'X-ray microscopy'.

It was to be a major source of inspiration to D. Gabor in his invention of holography (§5.4) and thence to the use of visible light in processing electron microscope images, developed by A. Klug and his colleagues (§5.5.1).

As mentioned above, the lack of experimentally determined X-ray phase data, required for image formation to be completed by optical reconstruction, prevents the method from being used for the routine determination of crystal structure. The solution to that problem — the 'phase problem' of X-ray crystallography — came mainly from the development of mathematical methods that enable phases to be deduced, with the aid of the computer, from X-ray intensity data and the fact that there are certain physical boundary conditions such as the finite sizes of atoms and the absence of negative electron density.

The value of the optical analogue continues to be the insight it provides into structure determination, and the special uses to which it can be put, such as the study of disorder and other effects in semi- and poly-crystalline materials. The modern techniques used in optical transform work (of which Figs. 5.06 and 5.10 are examples) have largely been developed by Lipson, Taylor and others in Manchester since the 1950s. The *Atlas of optical transforms* (Harburn *et al.*, 1975) contains a wide variety of illustrations.

(iii) Lens aperture and aberrations
In Fig. 5.10(a) (c) (e) circular apertures of diminishing size have been introduced in the centre of the diffraction plane to limit the amount of information allowed to contribute to the formation of the image. By removing the outer diffraction spots the masks increasingly deprive the image of the higher spatial frequencies. The results, shown in (b) (d) (f) are correspondingly deficient in fine detail.

To calculate this effect of the aperture we can argue that it has an aperture function that multiples the diffraction pattern of the object by unity inside its boundary and by zero outside. Since the image is the transform of that part of the pattern in the diffraction plane that contributes to the image, we can then write

$$\text{Image} = T[T(\text{object}) \times (\text{Ap.fn})] \tag{5.08}$$

where T denotes Fourier transformation, and the appropriate limits must be specified for the aperture function (Ap.fn).

The *equivalence* of this to the Rayleigh convolution treatment of imaging (§5.2) is shown by applying the convolution theorem to the above equation, which immediately gives

$$\text{Image} = \text{Object} \circledast T(\text{Ap.fn}) \tag{5.09}$$

i.e. the image is the convolution of the object distribution with the Airy pattern given by the lens aperture.

In accordance with the Rayleigh model each point of the object is here being regarded as a source and is being imaged as the Airy pattern of the lens aperture.

With regard to aberrations, their effect is to modify the aperture function. This has no effect on the principles involved above and for a treatment of aberrations reference should be made to the specialist books.

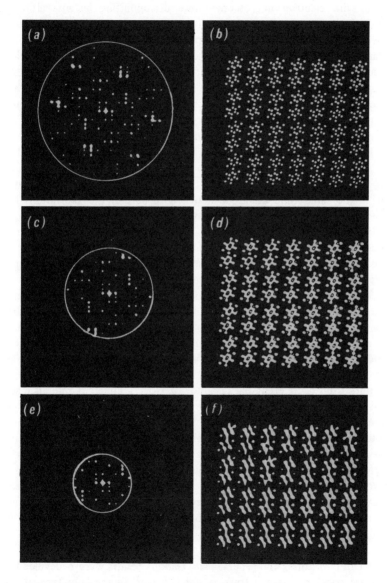

Fig. 5.10 – Deterioration of image quality as high spatial-frequencies are removed by filtering in the diffraction plane. (Reproduced by permission, from Taylor and Lipson (1964), *Optical transforms*. Bell, London.)

5.4 HOLOGRAPHY

The invention of holography in 1948 by D. Gabor, for which in 1971 he was awarded the Nobel Prize in physics, stemmed from his work on improving the quality of images obtained in electron microscopy. In the 1940s the results obtained with electron microscopes were disappointing because although a hundredfold improvement on the resolving power of the best light-microscope had been obtained, the resolution stopped far short of the theoretical value. The fast electrons used in electron microscopy have a de Broglie wavelength of about 1/20 Å, so that atoms should be resolved; but the practical limit at that time was approximately 12 Å. A major reason for the shortfall was the presence of aberrations associated with the electron lenses used. It was in thinking about how to solve that problem that Gabor devised the technique he called **wavefront reconstruction**. His inspiration came partly from the principles involved in W. L. Bragg's two-wavelength microscopy (§5.3.3). He reasoned that if he could record the phases as well as the intensities in an electron microscope image, then perhaps he could complete the image formation in an optical system which at the same time could be designed to correct the aberrations in the electron optics. The time-honoured way to record phases was to change them into interference effects and in thinking about how to do that in this particular instance, Gabor was influenced by Zernike's successful use of a 'coherent background' to investigate lens aberrations in such a way that phases and intensities were revealed. For the first stage then, the electron microscope would be used to produce an interference pattern between the scattering from the object and the unused 'background' (i.e. the non-scattered part of the illuminating electron beam).

Gabor's original experiments were solely with visible light and used a method that was soon superseded: we need not describe it here. The principle of the method, as opposed to the practical arrangement, is shown in Fig. 5.11. A single object point, B, is considered. Wavefronts scattered from B interfere with wavefronts from a coherent background, or from a coherent reference source such as A in the figure. Suppose that the interference pattern is photographically recorded as a positive which, after processing, will only transmit at the positions of the interference maxima. Then, if the photograph on its own is now illuminated by the reference source, the wavefronts transmitted by it are still in phase with the reference source. But as interference maxima they are also in step with wavefronts that would leave B if B were still there. Thus wavefronts from B could, despite its absence, be said to have been 'reconstructed'. The photograph has acted as what Gabor called a 'hologram' − from the Greek word *holos* meaning whole − to indicate that it contains the whole of the information about the object, both amplitude and phase.

Though confirming the general idea, Gabor's results were bedevilled by the inadequate coherence length (only about 0.1 mm) of the light from the high-pressure mercury vapour lamp that was used, and by the low level of the illumination available after the introduction of a small pinhole (3 μm diameter) to secure

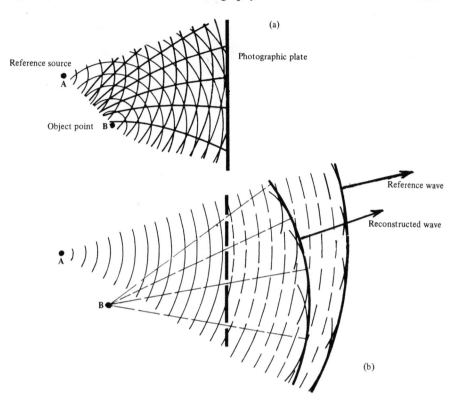

Fig. 5.11 – Gabor's schematic portrayal of wavefront reconstruction: (a) recording, (b) reconstruction (Gabor, 1972).

adequate spatial coherence. For these and a number of other reasons, the envisaged application to electron microscopy was unsuccessful. As Gabor described it, holography went into a long hibernation. The revival came with the work of E. N. Leith and J. Upatnieks (1962). Their success was due to their recognition of the similarity of Gabor's wavefront reconstruction process to the theoretical work that Leith and his collaborators had been undertaking on 'side-looking' radar. It involved the use of 'skew' reference illumination and led to a considerable improvement (Leith and Upatnieks, 1963, 1964). Use of the recently developed lasers in this work followed (Leith and Upatnieks, 1965) and the combination of these two advances led to a more generalized and improved process of holography.

One of the geometries for recording a 'Leith–Upatnieks hologram' is shown in Fig. 5.12(a). Coherent plane-wave illumination is scattered by (in this example) a transparent object, and the hologram is formed by making that scattered illumination interfere with a reference beam produced by suitably deflecting an unused part of the incident illumination. To appreciate how the hologram

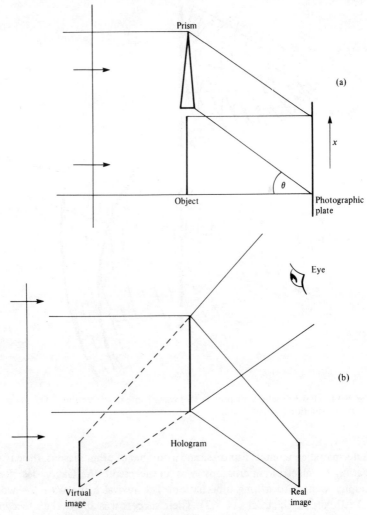

Fig. 5.12 — The Leith–Upatnieks method for holography.
(a) Recording. (A prism is commonly depicted in this type of diagram, though a mirror is usually used to deflect the reference beam in practice.)
(b) Reconstruction.

obtained by photographically recording this interference pattern carries the amplitude and phase data required for 'reconstructing' an image of the object, we need only consider the process in one dimension, the x-axis in the figure.

Let A_x and δ_x be the amplitude and phase, respectively, of the illumination at x, that has arrived from the object. The amplitude, A_0, of the reference illumination at the hologram can be assumed constant, and its phase at any point

x is $(2\pi/\lambda)x \sin\theta = 2\pi\alpha x$ say, where θ is the angle of incidence of the reference beam and $\alpha = (\sin\theta)/\lambda$. The total amplitude, R_x, as a function of x, can therefore be expressed as

$$R_x = A_0 e^{i2\pi\alpha x} + A_x e^{i\delta x} \ . \tag{5.10}$$

The intensity I_x at x is then given by

$$I_x = |R_x|^2 = A_0^2 + A_x^2 + A_0 A_x e^{-i(2\pi\alpha x - \delta_x)} + A_0 A_x e^{i(2\pi\alpha x - \delta_x)} \tag{5.11}$$

$$= A_0^2 + A_x^2 + 2A_0 A_x \cos(2\pi\alpha x - \delta_x) \tag{5.12}$$

The cosine term represents a fringe system recorded in the hologram. It carries (hence *'carrier fringes'*) all the information about the object wavefront system because the amplitude, $A_0 A_x$, of the fringes is proportional to A_x (A_0 is constant) and the position of the fringes (spacing $1/\alpha$) with respect to x is determined by δ_x.

Used for reconstruction, the hologram can be illuminated with parallel coherent light at normal incidence (Fig. 5.12(b)). By a suitable photographic development process the amplitude transmission factor in the hologram can be linearly related to I_x in Eqns (5.11) and (5.12). The *amplitude*, $(A_x)_{rec}$, of the reconstructed wavefront is therefore given directly by the right-hand sides of those equations if we omit a multiplying factor that does not affect the interpretation. Using Eqn (5.11) we have

$$(A_x)_{rec} = A_0^2 + A_x^2 + A_0 A_x e^{-i(2\pi\alpha x - \delta_x)} + A_0 A_x e^{i(2\pi\alpha x - \delta_x)} \ . \tag{5.13}$$

The first two terms represent the illumination in the direction of the reconstruction beam. This is essentially A_0^2, as A_x^2 is normally arranged to be much less than A_0^2. The third and fourth terms can be rewritten as follows

$$A_0 e^{-i2\pi\alpha x} A_x e^{i\delta x} + A_0 e^{i2\pi\alpha x} A_x e^{-i\delta x}$$

respectively. In the first of these, $A_x e^{i\delta x}$ is exactly the complex amplitude wavefront distribution that originally left the object to form the hologram. Alone, it would enable a (virtual) image to be formed of the object in its original position. However, multiplication by $A_0 e^{-i2\pi\alpha x}$ imposes a phase shift which amounts to a rotation, making it necessary to view the hologram in direction $-\theta$ (Fig. 5.12(b)) in order to see the image. In the second term, $A_x e^{-i\delta x}$ corresponds to a (real) image of the complex conjugate of the object, and multiplication by $A_0 e^{i2\pi\alpha x}$ means that it has to be viewed from direction θ. Being the complex conjugate of the object (cf. the negative sign in the exponential) this latter image is inverted in a way that makes it appear to be inside out — technically referred to as *pseudoscopic*.

It will be noticed that in the above simple account of reconstruction, light at normal incidence has been used, not a duplicate of the original reference beam. When the thickness of the recording emulsion is taken into account, however, the conditions do become more critical.

Other arrangements are now commonly used for producing holograms of opaque and transparent objects, for 3-dimensional colour imaging, and for the various applications in microwave technology, acoustics, 'non-coherent photography', non-destructive testing, motion studies, information storage, etc. Descriptions are to be found in most modern mainstream physics textbooks. Many of the useful properties of holograms are not associated, however, with these developments in the techniques for imaging, so we shall not describe them here but will restrict ourselves to some Fourier aspects of holography in §5.5.1(iii).

5.5 OPTICAL PROCESSING

5.5.1 Coherent processing

In describing the two-stage process of coherent optical image formation we have seen that the quality of an image is determined by the information reaching it from the object. Specifically, the quality depends on the faithfulness with which its constituent spatial frequency spectrum is a replica of that of the object. We have also seen that the spatial frequency spectrum is accessible in the diffraction plane. Figure 5.10 showed an example, one in which image quality deteriorates when high frequencies are removed by simply placing an aperture in the diffraction plane to prevent frequencies above a certain value from contributing to the formation of the image − a *low-pass filter*.

That was a very simple example of coherent **spatial filtering**, which now we need to consider in more general terms, because there are other types and methods of filtering in the frequency domain whose range of applications is far-reaching.

The basic arrangement is restated in simplified form in Fig. 5.13. The object is in the form of a transparency that transmits a complex amplitude $f(x)$ − keeping

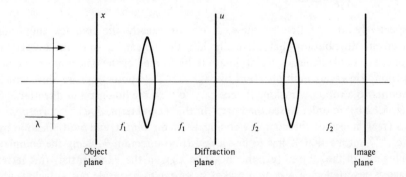

Fig. 5.13 − Basic arrangement for coherent optical filtering.

to one dimension solely for simplicity in expressing the mathematics. The transparency is illuminated at normal incidence with light from, for example, a laser and beam-expander combination to produce planewave coherent illumination. In the diffraction plane we have the Fourier transform of the transparency,

$$F(u) = \int f(x) \, e^{-2\pi i u x} \, dx \quad . \tag{5.14}$$

The image is formed as a further stage of Fourier transformation.

The opportunity for spatially filtering the image is in the diffraction plane and the example (Fig. 5.10) at the beginning of this section is one of *'amplitude filtering'*. (In particular, it is an example of a *'blocking filter'* because of its all-or-nothing nature at any given frequency.)

Filtering with respect to phase, or a combination of amplitude and phase — *complex filtering* — is also practicable.

Optical filtering, then, is a particular form of optical processing in which the spatial Fourier transform of an object is operated on in such a way as to have a predetermined effect on the image. It has its origins in the Abbe theory of image formation.

Examples of the three types of filtering will help to give some idea of their range of application.

(i) Amplitude filters

A valuable application of this type of filtering was introduced in 1964 by A. Klug and J. E. Berger for the analysis of electron micrographs of large biological molecules by optical diffraction, using the electron micrograph as a diffraction mask.

In many biological specimens there is a component of structural regularity that is of interest, for example the arrangement of protein sub-units in virus particles. Often, however, the details are not clearly revealed in the electron micrographs because of the presence of, for example, non-organized material in the specimen. The procedure developed by Klug and his collaborators can be described by reference to Fig. 5.13. The electron micrograph is placed in the object plane, and the pattern it gives in the optical diffraction plane is photographed. Only the component in the object that has structural regularity gives a regular diffraction pattern. Other diffraction effects, in this case arising from the non-organized material, can be recognized and a mask made to withold them. With the electron micrograph still in the object plane, and the filtering mask now in the diffraction plane to allow only the chosen diffracted beams to proceed, an image is obtained in which there is improved clarity in the imaging of the periodic component of the object.

Another example is illustrated in Fig. 5.14. In (a) we have an electron micrograph of a tubular structure found in a form of bacteriophage. The outer coating of the tube is composed of protein molecules of molecular weight

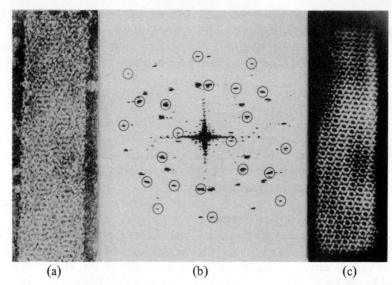

Fig. 5.14 – Optical filtering in electron microscopy.
(a) Electron micrograph of negatively stained flattened tubular structure in a bacteriophage (× 200 000).
(b) Optical diffraction pattern of (a). The circles are round diffraction spots assocated with the structure of one side of the flattened tube. A mask is made to allow only these to form the image.
(c) Resulting filtered image, showing helical structure consisting of molecules arranged in hexamers.
(DeRosier, D. J., and Klug, A. (1972), *J. Mol. Biol.*, **65**, 469–488.)

approximately 50,000 and arranged in a helix like a coil spring. To obtain this picture the specimen had first been embedded in an electron-dense medium (uranyl formate). This is a standard procedure. The medium scatters electrons more efficiently than the protein, and by occupying the holes and crevices in the surface it throws the surface structure into relief. However, tubular specimens like this become flattened when prepared for electron microscopy, and the details of the back and front of the tube become superimposed. In (b) and (c) we can see how the images can be separated. The optical diffraction pattern of (a) is shown in (b): it was obtained in the same way as for the example described above. Circles have been drawn round the diffraction spots that could be identified as corresponding to the structure of one layer (the back, say) of the tube. With a mask that allowed only these to proceed to the image plane the picture shown in (c) was obtained. The arrangement of the individual molecules (in hexamers) can now be seen clearly. Of course, great care has to be exercised in this work since if an unrelated set of diffraction spots were selected a false image would be produced.

The same principles are now used for processing electron micrographs by computer. The photographic image is converted into digital form by densitometry and a computer is used to perform the Fourier transforms and the filtering. By using this method phases as well as intensities are available and the procedure is generally more amenable than the optical method for correcting for aberrations and other unwanted effects due to the electron optics of the microscope. Regarding an electron micrograph as an aperture function, albeit a very complicated one, its Fourier transform can be calculated in full detail of amplitude and phase. (Because they are not 'seen', phases are all too readily forgotten in optical processing in general; and in work such as the above they can be as important as, or more important than, the amplitudes. As we have noted elsewhere, however, optical methods do have their advantages.)

A further major advance by Klug and his collaborators has been the development of a technique for producing well-resolved 3-dimensional images of large biological molecules (DeRosier and Klug, 1968). Figure 5.15 shows an example in which a 3-dimensional image of human wart virus has been 'reconstructed'. Calculated by combining the Fourier terms of a number of ordinary 2-dimensional pictures photographed in the electron microscope at different angles of tilt of the specimen, it provides a much clearer image than is obtained in any one of the original 2-dimensional views alone. This is because it eliminates the confusion present in the individual pictures due to the overlap of features lying within the depth of focus of the electron microscope objective. Prior to this development the latter had been considered to be a limitation inherent in electron microscopy.

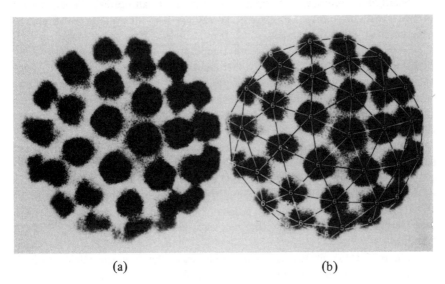

(a)   (b)

Fig. 5.15 – (a) A 3-dimensional reconstructed image of human wart virus (c. 500 Å dia.),
(b) shows the underlying icosahedral surface lattice with the 5-fold and 6-fold vertices marked ($1\text{Å} = 10^{-10}\text{m}$).
(Klug, A. and Finch, J. T. (1968), *J. Mol. Biol.*, **31**, 1–12).

For this work Klug was awarded the Nobel Prize in chemistry in 1982.

3-dimensional imaging in this way is an extension of the conceptual framework of X-ray crystallography. The main difference is that in electron microscopy amplitudes and phases can be calculated because, as noted above, an image is available, whereas in X-ray crystallography only X-ray diffraction intensities can be measured.

Much the same principles underlie the new techniques of X-ray computerized tomography ('X-ray scanning'), positron emission tomography and NMR zeugmatography (for which see Pykett (1982) for a simple exposition).

Insofar as we have mentioned the importance of phase and the need for phase data, we have trespassed somewhat on the following paragraphs. Other applications of *amplitude* filtering include the removal of raster from TV pictures, removal of the dot structure of half-tone pictures (this is the 2-dimensional equivalent of raster removal), removal of additive noise, and image contrast control. The last of these is done by altering the balance of high- and low-frequency contributions to images.

(ii) Phase filters

We have referred in (i) to the importance of, and the frequent need for, phase data. An application of filtering in which phases play the leading role has been in the field of optical microscopy.

Sections of biological materials examined under an optical microscope are often almost, or totally, transparent. To say the least this makes it difficult to observe their structure, unless some trick is used. With such materials it is their refractive index, and therefore their optical thickness, that varies from point to point. Because this only introduces phase differences between light that has passed through different regions it has no effect on the amplitude of the transmitted light and so is invisible to the human eye. For obvious reasons, materials of this type are referred to as *phase objects* as opposed to *amplitude objects* with which we have been mainly concerned.

The classical way of revealing the structure of materials like this is to use selective chemical staining, and the value of that method cannot be overemphasized. A more direct method, however, for distinguishing between regions of different refractive index (and therefore of different biological composition) is by **phase-contrast microscopy**, originally proposed by F. Zernike in c. 1935 (see Zernike, 1942), for which he was awarded the Nobel Prize in 1953 for physics.

Figure 5.16(a) illustrates, in a simple way, the basic principle involved. The phasors labelled A represent the illumination at two points in the image plane of a phase object. They have the same amplitude, A, but their phases differ because of differences in the optical thickness of the object. If illumination of constant amplitude and phase is superimposed across the field, as represented by the dotted phasors labelled B, interference occurs and the resultants, R,

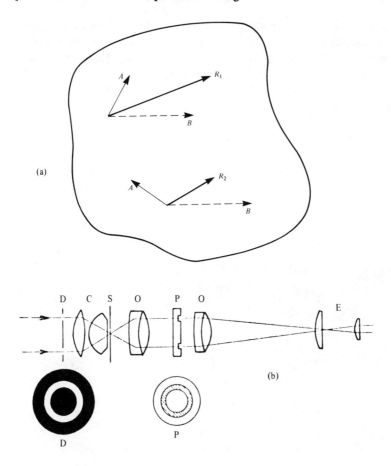

Fig. 5.16 – Phase-contrast microscopy.
D, diaphragm; C, condenser; S, specimen; O, components of objective; P, phase plate; E, eyepiece.
(After Barer, 1955)

are different in amplitude as well as phase. The optical path differences have therefore been made 'visible'. One of the simplest arrangements for doing this is shown schematically in Fig. 5.16(b). Light is focused on the specimen S by a condenser lens C. A diaphragm D in the form of an annular slot is placed in the focal plane of the condenser, and its image is formed as a ring of light in the back focal plane (P) of the objective O by light that has not been diffracted by the specimen. The effects illustrated in (a) can be achieved, therefore, by introducing a phase difference between this 'reference' illumination and the light diffracted by the specimen. (The small proportion of the *diffracted* light that passes through the ring in the back focal plane simply dilutes the effect.) The phase difference is introduced by means of a **phase plate** at P consisting of an optically parallel glass plate on which a thin layer of dielectric material has been deposited over the area of the ring, or everywhere except the ring, giving 'negative' or 'positive' phase contrast, respectively. The example shown in Fig. 5.17 is of squamous cells from the lining of the mouth. Details of the cell architecture are more clearly seen with phase contrast (a) than with normal bright-field illumination (b).

More complicated methods of what can collectively be called **interference microscopy** have been developed since Zernike's pioneering work. They generally involve some form of beam splitter and the controlled introduction of a phase difference in one beam before recombination. The Nomarski differential–interference–contrast microscope is particularly popular at the moment and is well described and referenced by Padawer (1968).

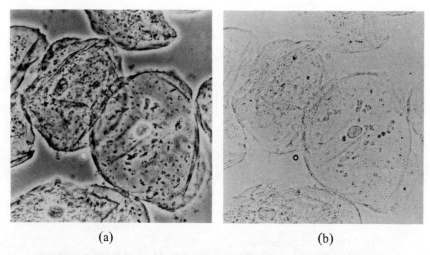

(a)   (b)

Fig. 5.17 – An example of phase-contrast microscopy. Optical micrographs of squamous cells from the lining of the mouth. (Field of view c. 100 μm.)
(a) Phase contrast. (b) Conventional bright-field illumination.
(Courtesy of Dr Martin C. Steward, Medical School, University of Manchester.)

(iii) Complex (holographic) filters
With its ability to record phase, holography has led to considerable advances in the use of filtering.

The idea of holographic filters was first seriously mooted by A. Vander Lugt in 1963 (his 1964 paper is more accessible) in connection with their possible use in signal detection. Since then their use has been extended to include correction ('balancing') of aberrations in optical systems, image-motion compensation, etc. Before considering applications we need to know the basis on which this type of filtering operates.

[§5.5] Optical Processing

The type of hologram we have already considered is described as either a Fresnel or a Fraunhofer hologram depending on whether the photographic plate used for recording it is located in the near- or far-field. Such holograms are used in 'lensless' photography of 3-dimensional objects and for many other applications of holography. A hologram can be formed in any plane, however, and with the arrangement in Fig. 5.18 it is recorded in the focal plane of the first lens, the Fourier transform plane of the object. This **Fourier transform (or generalized) hologram** has properties of special value in certain types or filtering. (There is also a 'lensless' geometry for recording this type of hologram.)

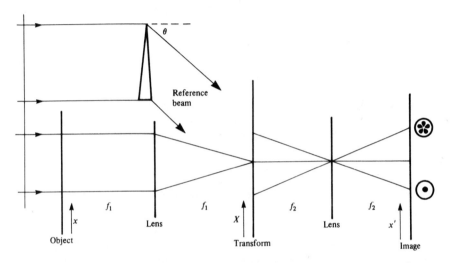

Fig. 5.18 – Coherent optical processing. Optical correlator using a matched spatial-filter for pattern recognition.

Restricting this treatment to one dimension for clarity, the x-axis as indicated in Fig. 5.18, let f_x be the complex amplitude transmitted by the coherently illuminated object. In the focal plane of the first lens we have the Fourier transform of f_x, which we can denote by F_X. The capital F represents the transform of f in the conventional way and X denotes the coordinate in the Fourier plane. (To allow us to see the relationships as clearly as possible we omit various factors such as the wavelength of the illumination, focal lengths of lenses, etc., which only affect the geometrical aspects.)

Let $a_0 e^{-i2\pi\alpha X}$ be the complex amplitude of the reference illumination arriving at the transform plane, where $\alpha = (\sin\theta)/\lambda$ as in §5.4.

The intensity distribution, I_X, in the transform plane is given by

$$I_X = |a_0 e^{-2\pi i\alpha X} + F_X|^2$$
$$= a_0^2 + |F_X|^2 + a_0 F_X e^{2\pi i\alpha X} + a_0 F_X^* e^{-2\pi i\alpha X} . \quad (5.15)$$

This is photographically recorded as a transparency as described earlier, so that the hologram has an amplitude transmittance proportional to this intensity distribution.

Now consider what happens when this hologram is replaced in the transfrom plane, the reference illumination shut off, and a different object used, transmitting with a different pattern of complex amplitude, a_x say. In the Fourier plane the latter is A_X. 'Filtered' in the Fourier plane by the hologram whose complex amplitude transmittance is given by Eqn (5.15), the amplitude leaving the transform plane, U_X say, is therefore

$$U_X = a_0^2 A_X + |F_X|^2 A_X + a_0 F_X A_X e^{2\pi i \alpha X} + a_0 F_X^* A_X e^{-2\pi i \alpha X} \ . \quad (5.16)$$

In the focal plane of the second lens we have an image, of complex amplitude $\psi_{x'}$ say, which is the Fourier transform of U_X. Using the convolution theorem and the transform of a δ function (Eqn (4.20)) freely, gives

$$\psi_{x'} = a_0^2 a_{x'} + (f_{x'} \circledast f_{x'} \circledast a_{x'})$$
$$+ [(a_0 f_{x'}) \circledast a_{x'} \circledast \delta(x' + \alpha)]$$
$$+ [(a_0 f_{-x'}^*) \circledast a_{x'} \circledast \delta(x' - \alpha)] \ . \quad (5.17)$$

(Concerning the negative sign in the subscript of $f_{-x'}^*$ see the footnote in Eqn (4.51) (p. 81).)

The first two terms refer to illumination centred on the axis in the image plane and is not normally of any use. The third term is the convolution of $f_{x'}$ and $a_{x'}$, and is shifted by the δ function to be centred at $x' = -\alpha$: we will call this the **convolution image**. The last term is likewise the convolution of $f_{-x'}^*$ and $a_{x'}$ and is centred at $x' = \alpha$. It is also the cross-correlation of $f_{+x'}^*$ and $a_{x'}$ (see Eqn (4.46)) and is referred to as the **cross-correlation image**.

Note that the above findings are an example of how multiplication in the Fourier (diffraction) plane corresponds to convolution in the image plane.

The hologram of f_x above is referred to as a **Vander Lugt filter** and the following examples will illustrate the uses of the two displaced images.

THE CROSS-CORRELATION IMAGE
Example: pattern recognition (character recognition).

Suppose that one wishes to identify a particular 'pattern' in a series of 'object' transparencies. A Fourier transform hologram filter of the particular pattern, such as our f_x above, is made in the manner we have described and replaced in the transform plane. The identity parade of object patterns being inspected are located in turn in the object plane in Fig. 5.18. When it is the turn of the pattern being sought the cross-correlation image will have a bright central spot.

All patterns which do not match f_x will give zero illumination in that image (cf. §4.7) — unless there is some similarity, when a weak 'parasitic response' will be obtained there.

Filters of this type, comprising the Fourier transform of a sought pattern, are collectively known as *'matched filters'*.

THE CONVOLUTION IMAGE
Example: deblurring, aberration compensation.

From earlier chapters it will be evident that the effect of aberrations, or any other feature of an imaging system that results in a blurring of the image, can be regarded as a convolution of the ideal image, a_x say, with a blurring function, g_x (cf. §5.3.3(iii)). The normally produced optical image would therefore be the convolution $a_x \circledast g_x$. To remove the effect of g_x one can therefore multiply (*vide* the convolution theorem) the transform of this convolution $(A_X G_X)$ at the diffraction stage, by $1/G_X$. To make a filter to do this, light from a point source is passed through the aberrated lens system and interfered with a collimated reference beam. There are snags, however, and the reader should consult the specialist books for details.

•   •   •   •   •

A disadvantage of processing under coherent conditions is that the handling of complex amplitudes precludes the use of TV displays or LED arrays as inputs to the processing system. Furthermore, coherent processing tends to be troubled by 'noise' produced by dust, scratches and other blemishes on optical components. Also, when the output from such systems is presented it is usually in the form of intensities, so that phase data are lost.

5.5.2 Incoherent processing

Because of the disadvantages inherent in coherent processing mentioned in the previous section, a trend during the past decade has been towards incoherent processing. Needless to say, there are disadvantages in that too and choice must depend on the nature of the problem to be handled. With incoherent processing, inputs and outputs normally need to be non-negative and real. Bipolar or complex-valued processing can be performed, however, by multiplexing — but then some form of hybrid system is usually required.

Two main kinds of incoherent processing can be identified and these are briefly described below, but only in the simplest outline because they refer to topics that are really outside the subject of this book. The first is again concerned with the use of diffraction, the other is based on purely geometrical principles in which diffraction plays no active role.

(i) Diffraction-based processing
Example: pattern recognition by energy-spectrum correlation.

With the arrangement shown in Fig. 5.19 a transparency T_1 is illuminated with coherent plane-wave quasimonochromatic light. Its transmitted complex amplitude can be denoted by f_x and its Fourier transform in the focal plane of lens L_1 by F_X, using 1-dimensional representation again for simplicity. A diffuser in the transform plane destroys the coherence there and produces an incoherent, effectively 'self-luminous', intensity distribution proportional to $|F_X|^2$ which is the energy spectrum of f_x (cf. §4.7.1).

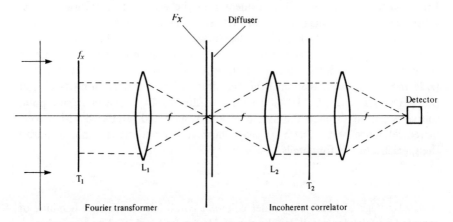

Fig. 5.19 – Incoherent optical processing. Energy-spectrum correlator.

Another transparency is located at T_2. The output from the system is therefore the convolution of $|F_X|^2$ with the energy spectrum of the transparency at T_2. Since T_1 and T_2 are normally real, if either is inverted the convolution is equivalent to correlation. The instrumentation represented schematically in the figure is commonly known as an **energy-spectrum correlator**.

Used for alphanumeric pattern recognition, for example, the technique depends on the fact that only if the two transparencies T_1 and T_2 have the same spatial frequency spectra is the cross-correlation output from the system (usually measured with a photo-detector on the axis) significantly non-zero. A transparency of the 'required' pattern is used for T_2 and the unknowns to be searched are located at T_1. Success depends upon the various spectra all being different. Against this shortcoming the method has a distinct advantage over that described in §5.5.1 in that the measured cross-correlation output is independent of the position and orientation of the patterns in the input plane because their spectra are not directional.

Applications include alphanumeric character recognition in language translation, information retrieval, etc.

(ii) Geometrical optics-based processing
Example: pattern recognition.

In Fig. 5.20 light from a source S is collimated by lens L_1 and passes through a photographic transparency T_1 and then through a second transparency T_2 that can be displaced in its own plane. Lens L_2 focuses the light transmitted through T_1 and T_2 on to a photo-detector at its focal point.

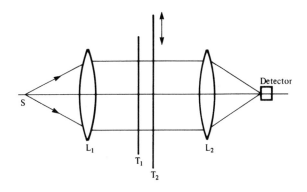

Fig. 5.20 – Geometrical optics-based processing. T_1, T_2. transparencies in x,y plane perpendicular to the figure.

Let the intensity transmittance of T_1 and T_2 be $t_1(x, y)$ and $t_2(x + u, y + v)$ respectively, where u, v is the translation of T_2 with respect to T_1.

The detector integrates the light transmitted by the two transparencies, and its output, P, a function of u, v is given by

$$P(u, v) = \iint t_1(x, y) \, t_2(x + u, y + v) \, dx \, dy \ . \tag{5.18}$$

This is the cross-correlation of $t_1(x)$ and $t_2(x)$ (cf. Eqn (4.40)) and from §4.7 it will be obvious that in general a strong signal is detected only if the two patterns comprising f and g are the same and if $u = v = 0$.

This *optical multiplier and integrator* can therefore be used for pattern recognition though, unlike the energy-spectrum correlator in (i), it requires that for a pattern T_2 to be identified with a pattern T_1 the two must both be in the correct rotational orientation.

This brief note is only intended as a hint at the types of processing that are possible with geometry-based incoherent imaging and related systems. They include spatial and temporal scanning methods, multichannel processing, etc. and are designed around a wide variety of devices (LEDs, charge-coupled device (CCD) arrays, etc.).

● ● ● ● ●

Articles on applications and developments in the field of data processing are to be found in volumes specially dedicated to this field: they appear, for example, in the series published by Springer-Verlag under the general titles *Topics in Applied Physics* and *Topics in Current Physics,* and in the series *Progress in Optics* edited by E. Wolf and published by North-Holland, Amsterdam.

6

Interferometry and radiation sources

> Could we employ the ocean as a lens, and force truth from the sky, even then I think there would be much more beyond.
>
> Richard Jefferies (1883) *The story of my heart*

6.1 INTRODUCTION

We turn now from the use of radiation for structural studies, image formation and image processing of various types, to the complementary topic of studying the spatial and spectral 'structure' of radiation sources.

In Chapter 1 we have seen in very general terms how the visibility of interference fringes can reveal information about radiation sources. The two types of interferometry pioneered by Michelson are particularly sensitive for such investigations.

Initially, Michelson used his stellar interferometer to measure star diameters, but he foresaw how fringe-visibility measurements might provide information about the brightness distribution of a source and he illustrated this with simple examples.

With his spectral interferometer, Michelson measured the wavelength of spectral lines, but again he foresaw, and demonstrated in simple cases, its greater potential, this time for using fringe visibility to obtain detailed information about the fine-structure of spectra. Later he used it to standardize the metre in terms of the red line of cadmium.

In 1907 Michelson was awarded the Nobel Prize in physics for the development of these optical instruments and for the spectroscopic and metrological investigations which he carried out by means of them.

Today, the two techniques are seen as the archetypes of important methods in modern astronomy and spectroscopy. The principles involved are as important today as they were then. For that reason they are recalled in §6.2 and §6.3. They also provide a valuable background for an introduction in §6.4 to the modern interpretations based on the theory of partial coherence.

Fourier transform spectroscopy is briefly described in §6.5, and in §6.6 some of the main Fourier aspects of imaging in astronomy, by interferometry, are outlined.

6.2 MICHELSON'S STELLAR INTERFEROMETER

6.2.1 Introduction

To introduce the use of interference effects as an indirect way of using a telescope to measure the angular dimensions of astronomical objects, consider Fig. 6.01(a). This depicts an aperture screen containing two slits perpendicular to the figure, and located in front of the lens of a telescope (the same can be done with a reflecting telescope). Wavefronts arrive from all points across a star of *angular diameter* ϕ_0 (the angle it subtends at the earth). Only the extreme wavefronts are indicated in the figure, W_1 originating from one edge of the star disc, W_2 from the opposite edge. In the focal plane of the lens there is a continuous spread of $\cos^2$ intensity fringes (the source being incoherent), ranging from those shown due to W_1 to those due to W_2. The nett result is a fringe pattern, shown in (b), of visibility < 1. Note that the spacing of the fringes stays the same as if the source were a point, viz. $\Delta = f\lambda/D$ (Eqn 1.11).

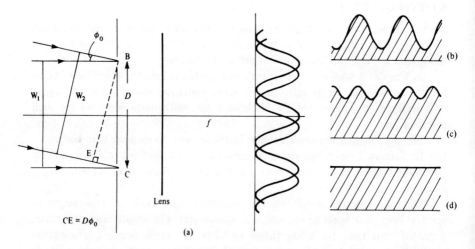

Fig. 6.01 — Measurement of stellar diameters by interferometry.

In practice the intensity of the fringe pattern is attenuated either side of the axis (cf. the sampling of the single-slit diffraction pattern in §2.4.1). We can ignore this attenuation if the slits are narrow, and particularly if, as in practice, observations are restricted to the central region of the fringe pattern.

Now if D is increased the fringe spacing is decreased and the fringe visibility is reduced, as indicated in (c). When D reaches a value such that the fringe spacing becomes just equal to the spread of the fringes, the visibility becomes zero, as in (d). The fringe spacing is now such that the fringes due to W_2 are displaced by exactly one fringe relative to those due to W_1. In diagram (a) the edge of the star sending wavefronts W_1 is equidistant from B and C, so that if the star were a uniformly luminous rectangle the wavefronts W_2 from the other edge would travel one wavelength further in their journey to C compared with their journey to B. At this special value of $D = D'$ say, we have

$$D'\phi_0 = \lambda . \qquad (6.01)$$

If D is further increased, the visibility becomes non-zero again, until $D''\phi_0 = 2\lambda$ when it is zero once more, and so on. (These results are of course not peculiar to the special alignment of the W_1 wavefronts with respect to B and C used here for mathematical convenience.)

As a star is effectively a circular luminous disc a correction has to be applied to the above. The method for making this correction is similar to that for deriving the resolving power of a circular aperture (§2.3). The result is that the first disappearance of fringes is not achieved until $D'\phi_0 = 1.22\lambda$, so that the star diameter is given by

$$\phi_0 = \frac{1.22\lambda}{D'} \qquad (6.02)$$

or, as a rough guide for yellow light,

$$\phi_0 \approx \frac{0.15}{D'_{(\text{metres})}} \quad \text{seconds of arc} . \qquad (6.03)$$

The principle of this method, suggested by Fizeau in 1868, was used successfully by Michelson in 1890 to measure the diameters of the moons of the planet Jupiter, obtaining values of around 1 second of arc. For this purpose he used the 12-inch (c. 0.3 m) telescope at Mount Hamilton.

With stars, however, the question arose as to whether atmospheric disturbances and mechanical vibrations of the telescope would prevent measurements from being made because, unlike planets, their diameters were expected to be only a few hundredths of a second of arc. These fears were partially allayed by the successful observation of fringes with the 100-inch (c. 2.5 m) reflector at Mount Wilson. But it was also realized that a telescope large enough to accommodate the separation of apertures necessary for achieving the disappearance of the fringes with such small stellar diameters would need to be 50-ft (c. 15 m) or more in diameter; and with slits at that separation the fringes would become too close to be observable. Furthermore, the result achieved was little different from the resolving power of the same telescope used in the normal way (cf. §2.3).

To overcome these difficulties Michelson devised the stellar interferometer associated with his name. It consisted of a linear array of four mirrors (Fig. 6.02) each about 6 inches (c. 15 cm) in diameter. The outer pair (M_1, M_2) acted as 'receivers', with separation (what would now be called the **baseline**) adjustable up to 20-ft (c. 6 m). The inner pair (M_3, M_4) were fixed, and they directed the incoming signals through two slits into a telescope to form fringes in the usual

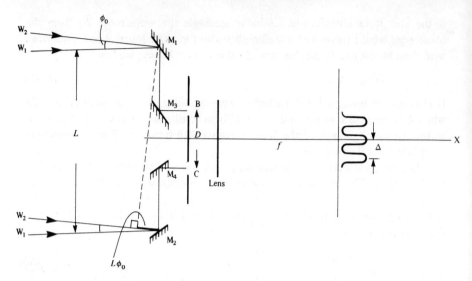

Fig. 6.02 – The Michelson stellar interferometer.

way. This scheme allowed the fringe spacing to be maintained at a constant and acceptable value ($\Delta = f\lambda/D$) determined by the separation (D) of the inner mirrors, whilst the path difference required for them to lose visibility was established when the separation L of the outer mirrors reached a value L' such that

$$L'\phi_0 = 1.22\lambda$$

i.e. $$\phi_0 = \frac{1.22\lambda}{L'} \quad . \tag{6.04}$$

The framework supporting the mirrors was built on to the 100-inch (c. 2.5 m) Mount Wilson reflecting telescope, now chosen not for its optical resolving power but for its ruggedness as a support for the frame carrying the mirrors. It was essential that they should be rigidly located with respect to each other since very small optical path differences through the system were being detected by the critical adjustment of the separation of the outer mirrors.

With this apparatus Michelson and F. G. Pease measured several giant and supergiant stars whose diameters were smaller than could be measured using the Mount Wilson telescope in the conventional way. For example, they obtained a value of 0.0047 secs of arc for Betelgeuse. (To allow for the darkening towards the 'limb' of a stellar disc they used data for the sun, and on that basis estimated that actual angular diameters would be about 17 per cent larger than values given by Eqn (6.04).)

In effect the new method gave a 'magnification' of L'/D compared with the first version, and limited only by L in theory. However, the difficulty of

[§6.2] Michelson's Stellar Interferometer

ensuring the necessary mechanical stability of a large baseline, together with the effects of turbulence in the atmosphere, prevented any significant further development. In 1930 Pease built a 50-ft beam version but it gave unreliable results and the work was abandoned in 1937.

Michelson realized, however, that the way in which the visibility changes as the separation of the outer mirrors is altered contains information not only about the size but also the brightness distribution of the source. Although the method could not be put to practical use at that time we will look at how he tackled the problem theoretically, because of the importance of the method today in astronomy.

6.2.2 Fringe visibility aspects

Michelson initially considered several examples of sources of different shape but with uniform brightness distribution (Michelson, 1890a). He used a construction essentially like that in Fig. 6.03 (with different symbols) and reasoned along the following lines.

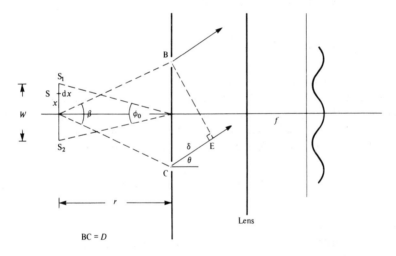

Fig. 6.03

An incoherent source of width W subtends an angle ϕ_0 at the interferometer which has slits perpendicular to the figure and whose separation D is adjustable. (D could equally refer to the separation of the outer mirrors in the later instrument.) Consider light from an elementary strip of width dx in the source at S, distance x from the axis. Let $y = f(x)$ be the length of the strip perpendicular to the figure. The illumination field from this strip is sampled at B and C, and the light diffracted from B and C gives a fringe pattern in the usual way. (The single-aperture 'envelope' of the fringe pattern can be ignored since only fringes very near to the axis need be considered.) To calculate the fringe intensity in direction θ we note that the path difference between the two routes SCE and SB is

$$= (SC + \delta) - SB$$
$$= (SC - SB) + \delta$$
$$= \beta x + \delta \quad .$$

((SC − SB) is evaluated in the same way as in Fig. 2.07, with i in that figure equal here to $\tfrac{1}{2}\beta$ and assumed small.) The resultant for unit amplitudes is found with the aid of a phasor diagram in the usual way to be

$$R(\theta) = 2 \cos \frac{\pi}{\lambda} (\beta x + \delta)$$

which gives the familiar $\cos^2$ type expression for intensity

$$R^2(\theta) = 4 \cos^2 \frac{\pi}{\lambda} (\beta x + \delta) \tag{6.05}$$

or
$$R^2(\theta) = 2 \left[1 + \cos \frac{2\pi}{\lambda} (\beta x + \delta)\right] \quad . \tag{6.06}$$

Since the source is incoherent each strip gives a set of separate fringes. And as the strips are of uniform brightness we can multiply the above expression by the area of the strip, $f(x)dx$, to represent the weighting of the contribution from that strip to the overall fringe pattern. In the limit when the strip width tends to zero the nett intensity, $I(\theta)$, in direction θ is given by integrating Eqn (6.06)

$$I(\theta) = \int_{-W/2}^{+W/2} f(x) \left[1 + \cos \frac{2\pi}{\lambda} (\beta x + \delta)\right] dx \tag{6.07}$$

ignoring the factor of 2.

To calculate the fringe visibility according to its usual definition (Eqn (1.05)) it is necessary to obtain the maximum and minimum values of intensity in the resulting pattern. One of the examples Michelson considered was a rectangular source of uniform brightness (Fig. 6.04(a)), whose width W was to be determined. The length of the source, $f(x)$, parallel to the slits, was now a constant and could be taken as unity for convenience. The integration in Eqn (6.07) then gives

$$I(\theta) = W + \frac{\lambda}{\pi \beta} \sin \frac{\pi \beta W}{\lambda} \cos \frac{2\pi \delta}{\lambda} \quad . \tag{6.08}$$

The maximum value of I occurs when $\delta = n\lambda$, where n is zero or an integer, and the minimum occurs when $\delta = (n \pm \tfrac{1}{2})\lambda$. Allowing for the gross distortion of scale in Fig. 6.03 we can write $r\phi_0 = W$ and $r\beta = D$, giving $\beta W = \phi_0 D$. The following expressions for I_{max} and I_{min} are then obtained

$$I_{max} = W + \frac{\lambda}{\pi\beta} \sin \frac{\pi\phi_0 D}{\lambda} \tag{6.09}$$

$$I_{min} = W - \frac{\lambda}{\pi\beta} \sin \frac{\pi\phi_0 D}{\lambda} . \tag{6.10}$$

Whence the fringe **visibility function**, $V(D)$, is given by

$$V(D) = \frac{I_{max} - I_{min}}{I_{max} + I_{min}} = \frac{\sin \frac{\pi\phi_0 D}{\lambda}}{\frac{\pi\phi_0 D}{\lambda}} . \tag{6.11}$$

(Denoting visibility as $V(D)$ expresses our interest in V as a function of D for a given λ and a fixed, unknown, value of ϕ_0.)

This is a sinc function (Fig. 6.04(b)) familiar in previous chapters as the diffraction pattern, and Fourier transform, of an aperture function having the same 'shape' as the brightness distribution in Fig. 6.04(a). The similarity between the two quite different examples is not a coincidence — but more of that presently (§6.4.1). The visibility curve in (b) drops to zero when $D = D' = \lambda/\phi_0$; and it is zero again when $D = D'' = 2\lambda/\phi_0$, and so on. This is in accordance with the interpretation in the previous section.

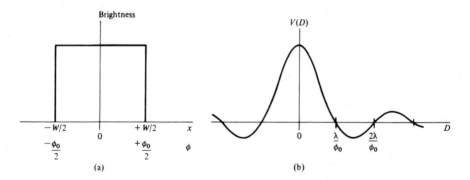

Fig. 6.04 — (a) Source brightness distribution. (b) Visibility curve.

Now to take non-uniformity of a brightness distribution into account. If we assume, with Michelson (1891a), that the total intensity from the elemental strip S at x in Fig. 6.03 is given by $B(x)$, this eliminates the added problem of the shape of the source. In effect, it reduces the problem to the determination of a

1-dimensional intensity distribution. Furthermore, if we consider the special case of $B(x)$ as an even function (a brightness distribution that is symmetrical about the axis in the figure) and if the variable x is replaced by ϕ as an angular variable (ranging from $-\phi_0/2$ to $+\phi_0/2$), the following expression for visibility is readily obtained

$$V(D) = \frac{\int B(\phi) \cos(2\pi D\phi/\lambda) \, d\phi}{\int B(\phi) \, d\phi} \qquad (6.12)$$

For the general 1-dimensional case when $B(\phi)$ is not even, though still working in terms of what Michelson called the 'total intensity of a strip of width $dx \ldots$', an equation of the following type is obtained

$$[V(D)]^2 = \frac{[\int B(\phi) \cos(2\pi D\phi/\lambda) \, d\phi]^2 + [\int B(\phi) \sin(2\pi D\phi/\lambda) \, d\phi]^2}{[\int B(\phi) \, d\phi]^2} \qquad (6.13)$$

Michelson concluded, prophetically, that from visibility measurements 'the distribution of luminous intensity of globular masses may be inferred, which would furnish a valuable clue to the distribution of temperature and density in gaseous nebulae'.

An analogous situation arose in Michelson's theoretical study of fringe visibility in relation to spectral distribution (§6.3.2). It was in that connection that Lord Rayleigh (1892) drew attention to the Fourier relationship between fringe visibility and (spectral) intensity distribution. The same applies here, and to conclude this section it will be interesting and informative to pursue the Fourier transform aspect a little further, from a modern viewpoint.

The numerator in Eqn (6.12) is the cosine Fourier transform of $B(\phi)$, with the denominator acting simply as a scale factor. We can readily see that the visibility for each value of the baseline D yields information about one particular Fourier component of the brightness distribution. This is clearly brought out by using the convolution theorem (§4.5). Expressing the observed fringe pattern for a given D as the convolution of $B(\phi)$ with the instrumental response, the convolution theorem tells us that the transform of this convolution is the product of the separate transforms. Now the transform of the instrumental response is a set of $\cos^2$ fringes which have a single spatial frequency determined by the value of D. It follows, therefore, that the transform of the observed fringe pattern for that value of D contains only information about just one harmonic component in the brightness distribution of the source.

The relationship between fringe visibility and coherence is dealt with in §6.4, but first we will look at the complementary topic of fringe visibility as a function of spectral — rather than spatial — distribution in a radiation source, this time using Michelson's spectral interferometer.

6.3 MICHELSON'S SPECTRAL INTERFEROMETER

6.3.1 Introduction

Unlike the stellar interferometer, the spectral interferometer is based on interference by division of amplitude (§ 1.4). The basic design was given by Michelson in 1881 in connection with an experiment to test whether the Earth moves relative to an 'aether'. For that purpose he went on to develop the instrument on a larger scale, in conjunction with E. W. Morley (the historical 'Michelson–Morley experiment'). But the basic arrangement was used for the measurement of spectral wavelengths (later to standardize the metre in terms of the red line of cadmium) and the study of spectral fine-structure. It is the spectroscopic applications that remain of significance today, indeed of considerably increased significance.

Figure 6.05(a) shows schematically the arrangement in one of the several early versions of the interferometer. Light from a source S (usually an extended source) is amplitude-divided at the 'half-silvered' rear surface of a glass plate, O, into two beams, one reflected and one transmitted. The reflected beam travels to the mirror M_1 and is then returned to be partially transmitted through O to the telescope T. Meanwhile, the other beam — the one initially transmitted through

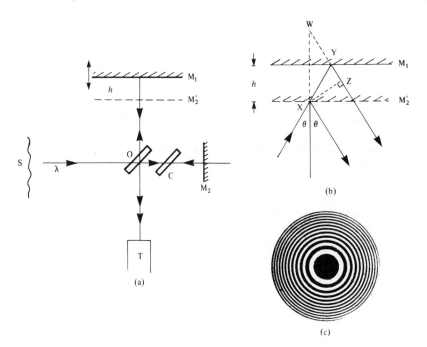

Fig 6.05 — The Michelson spectral interferometer.
    (a) Basic arrangement (refraction at glass plates O and C not shown).
    (b) Path difference between reflections is $XY + YZ = WZ = 2h\cos\theta$.
    (c) Appearance of the fringes with quasimonochromatic light.

the 'beam-splitter' – travels to mirror M_2 and is similarly returned to O, where it is partially reflected to the telescope. Since the beam sent to M_1 passes through plate O three times altogether, compared with the single passage of the beam sent to M_2, a 'compensating plate' of the same thickness and material as O is usually inserted at C. In general, M_1 and M_2 are not equidistant from O and there is an intentional path difference between the two routes (– the compensating plate is only used to equalize the dispersive paths through glass). Reunited, the two beams interfere, giving a resultant that depends on the path difference between them.

M_1 and M_2 are arranged to be mutually perpendicular, and the beam splitter is at 45° to both. The image of M_2 in O, as seen with the telescope, is at M_2', parallel to (or coincident with) M_1. The interference pattern observed with the telescope is therefore like that observed with the parallel plate in Fig. 1.08, though in the present example it is equivalent to being formed by reflection at an 'air plate'. With an extended source, of wavelength λ, rays can enter the system over a wide angular range and concentric bright rings are seen (Fig. 6.05(c)) (cf. Fig. 1.08(b)). Their circumferences are in directions θ for which there is reinforcement between the pairs of recombining wavetrains. That condition is given by

$$2h \cos \theta = m\lambda \tag{6.14}$$

where m is an integer or zero, and h is the mirror 'separation' (Fig. 6.05(b)). This assumes that the two interfering beams undergo identical phase changes at the beamsplitter. If that is not so, a constant has to be added to the phase difference associated with the path difference, and all fringes are displaced accordingly.

One mirror, M_1 in the figure, can be translated parallel to itself in the direction shown. Altering h causes the ring pattern to expand or contract; if h increases, the rings spread out from their centre as though generated there, and if h is decreased they shrink in to the centre.

An expression for the radial intensity distribution out from the centre of the pattern for a given value of h and wavelength λ is readily derived in the usual way with a phasor diagram. If the amplitudes of the light arriving at the telescope from the two routes are arranged to be equal, A say, then the resultant intensity, I_1, in direction θ in the ring system is found to be given by

$$I_1 = 4A^2 \cos^2(\psi/2) \tag{6.15}$$

where the phase difference

$$\psi = \frac{2\pi}{\lambda} 2h \cos \theta . \tag{6.16}$$

Therefore we have

$$I_1 = 4A^2 \cos^2\left(\frac{\pi}{\lambda} 2h \cos\theta\right) . \qquad (6.17)$$

The fringes for ideally monochromatic light are therefore of the $\cos^2$ type, as illustrated in Fig. 6.06(a). Furthermore, from what we have noted above about the effect on the ring pattern of altering h, it follows that if h is progressively increased or decreased, a detector at any point in the pattern (it could be on the axis, i.e. $\theta = 0$) would register a sinusoidal variation of intensity. If the light were indeed strictly monochromatic the wavetrains would be of infinite length (§4.6), and the sinusoidal pattern of fringe visibility would be quite unaffected by any amount of path difference introduced between the interfering light beams. Conversely, if such a pattern were experimentally observed one could deduce that the light was of such monochromaticity. If, in contrast, with another light source the visibility dropped to zero almost as soon as a path difference was introduced, it could be concluded that the light from the source had a broad continuous spectrum since the wavetrains must be very short (§4.6). It is the quantitative application of that philosophy to the analysis of optical spectra that is the key to the use of the interferometry method.

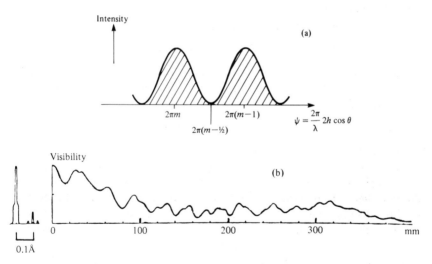

Fig. 6.06 – (a) $\cos^2$ fringes. (b) Michelson's results for Hg(λ 5461 A). (Adapted from Michelson, 1927.)

Consider for the moment a further hypothetical example. Suppose that the light being examined consists of two closely separated, ideally monochromatic wavelengths, λ_1 and λ_2. The changing pattern of intensity registered by our detector is now more complicated than in the above example of a single monochromatic wavelength. For the chosen θ-position of the detector there will be values of h for which rings of the two systems almost, if not totally and exactly, coincide: the detector will register a stronger signal. This happens when $h = h_1$ say, such that

$$2h_1 \cos \theta = p\lambda_1 \approx q\lambda_2$$

where p and q are integers. (In practice the two ring-systems can, for such a value of h, be almost in coincidence over quite a wide angular range if $\lambda_1 \sim \lambda_2$ is small.)

Increasing (or decreasing) h causes the two sets of rings to separate again, however slightly, and the detector records a succession of lower-intensity maxima and non-zero minima as they pass it. The details of the changing signal will depend on the difference between the two wavelengths, their relative intensities and, in real examples, on the line shapes and fine-structure. Since the two ring-systems move out (or in) from the centre of the pattern at different rates (cf. Eqn (6.14)) a new value of h is reached, $h = h_2$ say, for which 'coincidence' again occurs and the signal at the detector rises again; one set of rings has overtaken the other by one whole fringe interval. This condition can be expressed as

$$2h_2 \cos \theta = (p+k)\lambda_1 \approx (q+k+1)\lambda_2$$

where k is some number.

From the above equations we find, by subtraction, that

$$2(h_2 - h_1) \cos \theta = k\lambda_1 = (k+1)\lambda_2 \ .$$

This result can be used in various ways. For example, suppose that λ_1 is known and the observations are made on or near the axis, i.e. $\cos \theta \approx 1$. Put $2(h_2 - h_1) = H$. Then

$$\lambda_2 = H\lambda_1/(H + \lambda_1) \ .$$

Measurement of H therefore enables λ_2 to be calculated.

The considerable accuracy of this method for determining λ_2 is illustrated by a rough calculation that shows that the difference in wavelength between the two 'lines' in the sodium doublet (λ 589.0 nm and 589.6 nm) necessitates a mirror displacement of about 0.3 mm between successive coincidences. (We will not digress into the question of the resolving power as such, since our interest is really in the Fourier method described in §6.5.)

This way of using the interferometer is analogous to the earlier observation by Fizeau (1862) that by the 500th order in a Newton's rings experiment using a sodium source, the rings had almost disappeared (i.e. zero visibility), but that they regained their clarity at the 1000th order. He deduced that the sodium light was a doublet, with the 1000th order ring of the longer wavelength coinciding with the 1001th order ring of the shorter, the difference in wavelength between the two therefore being about 1/1000 of their mean.

Michelson's Spectral Interferometer

Michelson realized, however, that much information is lost in this method of analysis. He made visual estimates (transferred to a quantitative scale by a separate, ingenious calibration experiment) of fringe visibility as a function of mirror displacement. He realized that this 'visibility curve' contained very detailed information about the spectrum of the light source.

As early as 1887 Michelson had shown by careful observation and reasoning that 'the red hydrogen-line is a very close doublet; and the same is true for the green thallium-line'.

His mathematical analysis of these matters, together with an important contribution by Lord Rayleigh shortly afterwards in a published letter, are outlined in the following section because they provide a starting point for appreciating the basis of the Fourier transform method (§6.5).

6.3.2 Fringe visibility and spectral distribution

Michelson's research (1891b) under this heading is so (dare one say) illuminating and important in its modern relevance that it is helpful to have a clear picture of its content. The significant opening statement on the matter was: 'The general formula for the visibility of fringes due to interference of two streams of light whose difference of path is variable, from a source which is not homogeneous, is the same as that for a source of finite area whose "parallax" is variable'. (For 'homogeneous' we would say monochromatic.)

In other words, the suggestion was that the relationship between fringe visibility and spectral distribution is the same as that between fringe visibility and the spatial distribution of a source.

He started with the following expression for the intensity in direction θ in the fringe pattern obtained with the spectral interferometer, for light of wavelength λ:

$$I_1 = 4 \cos^2\left(\frac{\pi l}{\lambda}\right)$$

where l = path difference ($2h \cos \theta$ in Eqn (6.17)).

In spectroscopy the convention is to use wavenumber ($\sigma = 1/\lambda$) rather than wavelength. The above equation is then written as

$$I_1 = 4 \cos^2 \pi \sigma l . \tag{6.18}$$

If the light is polychromatic the fringe intensities in direction θ due to different wavenumbers are additive as there is no coherence between them. If a spectral intensity distribution $B(\sigma)$ extends from σ_1 to σ_2, the total intensity in direction θ is given by

$$I = \int_{\sigma_1}^{\sigma_2} I_1 \, d\sigma = 4 \int_{\sigma_1}^{\sigma_2} B(\sigma) \cos^2 \pi l \sigma \, d\sigma . \tag{6.19}$$

For σ substitute $\bar{\sigma} + \sigma'$ where $\bar{\sigma}$ is the mean wavenumber.

Replacing $B(\sigma)$ by $f(\sigma')$ then gives (omitting the numerical factor)

$$I = \int_{-a/2}^{+a/2} f(\sigma') \cos^2 \pi l \, (\bar{\sigma} + \sigma') \, d\sigma' \qquad (6.20)$$

where the spectral range from σ_1 to σ_2 is now expressed as being from $\sigma' = -a/2$ to $+a/2$ about the mean.

Expanding the $\cos^2$ term and following Michelson by putting

$$\left. \begin{array}{l} \int f(\sigma') \, d\sigma' = P, \quad \int f(\sigma') \cos 2\pi l \sigma' \, d\sigma' = C \\[6pt] 2\pi l \bar{\sigma} = \theta, \quad \int f(\sigma') \sin 2\pi l \sigma' \, d\sigma' = S \end{array} \right\} \qquad (6.21)$$

gives

$$I = P + C \cos \theta - S \sin \theta \; .$$

(Some changes in symbols have been made.)

If a is sufficiently small, the variations of C and S with θ may be neglected, and the maxima and minima of intensity occur when

$$\tan \theta = -S/C \qquad (6.22)$$

which gives $\quad I = P \pm \sqrt{C^2 + S^2}$

and therefore $\quad V^2 = (C^2 + S^2)/P^2 \; . \qquad (6.23)$

Michelson noted that this is of exactly the same form as the expression for visibility associated with a spatially extended source. Equation (6.13) referred to a brightness distribution $B(\phi)$, whereas here we have a spectral intensity distribution $f(\sigma')$.

It was in connection with Eqns (6.21) that Lord Rayleigh (1892) pointed to the Fourier relationship involved. At the same time he drew attention to the difficulty of using it to deduce a spectral distribution from visibility measurements because of the lack of information about the phases of the harmonics comprising the distribution: '. . . the visibility-curve by itself gives, not both C and S, but only $C^2 + S^2$; so that we must conclude that in general an indefinite variety of structures is consistent with a visibility-curve . . . '.

Despite this limitation, Michelson found it possible to deduce the structures of some simple spectral lines, and these were broadly confirmed by subsequent work.

The expressions for C and S in Eqns (6.21) are the coefficients in a Fourier representation of $f(\sigma')$ which Michelson wrote as

$$f(\sigma') = \int C \cos 2\pi l \sigma' \, dl + \int S \sin 2\pi l \sigma' \, dl \; .$$

Noting from (6.22) and (6.23) that $C = PV \cos \theta$ and $S = PV \sin \theta$ gave

$$f(\sigma') = \int V \cos(2\pi l \sigma' + \theta) \, dl \, . \tag{6.24}$$

If $f(\sigma')$ is symmetrical (remember that σ' is wavenumber relative to the mean value $\bar{\sigma}$), then $S = 0$ and

$$f(\sigma') = \int V \cos 2\pi l \sigma' \, dl$$

which allowed $f(\sigma')$ to be found by integration using the experimentally obtained visibility curve, V (or $V(D)$ as we denoted it earlier). In the case of an unsymmetrical $f(\sigma')$ it was necessary to know V and θ, the latter (which is also a function of l) being called the 'phase curve' by Michelson. In a number of examples he made estimates of θ by comparison with an 'approximately homogeneous symmetrical source of approximately the same wavelength'. (It is interesting to reflect that for this work Michelson developed his own mechanical sine and cosine analogue computer, which he called a 'harmonic analyser'.) Figure 6.06(b) shows an example of his investigations. It is the visibility curve for mercury (λ 5461 Å), and the deduced spectral distribution in which the two main components were estimated to be 0.1 Å apart (1 Å = 0.1 nm).

The development of the full potential of the method, however, had to await the advent of the modern digital computer. In its application for wavelength measurement the two-beam interferometer was superseded by the multiple-beam method used in the Fabry–Perot interferometer. Then in the 1950s it started to make its return, to become the basis of modern Fourier transform spectroscopy (§6.5).

6.4 PARTIAL COHERENCE, CORRELATION, AND VISIBILITY

The following treatment of partial coherence is essentially a simplified account of that given by Born and Wolf in their *Principles of Optics,* and largely based on the work of Zernike (1938) and Hopkins (1951). Unlike earlier studies, it is more closely related to experiment.

In Fig. 6.07 consider the illumination field produced in plane C by a source, W, that is both extended and polychromatic. Two points, C1 and C2, are chosen in the field, and by positioning a screen with a pinhole at each the visibility of interference fringes produced by the light from those particular places in the field can be experimentally measured. We seek the relationship that allows the temporal and spatial coherence of the illumination to be calculated from such measurements. After all, the fringe visibility is but a contrived physical manifestation of the coherence and we hope to relate it analytically to the more fundamental property.

Consider the interference at some general (off-axis) position of P. There is a time delay, τ say, between the journey times C1–P and C2–P. Without losing

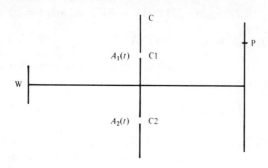

Fig. 6.07

generality this difference can be assigned (solely to simplify the equations we write) to one path. If the complex amplitudes of the illumination at C1 and C2 are $A_1(t)$ and $A_2(t)$ respectively, the intensity at P can be expressed as

$$I_P = \langle [A_1(t+\tau) + A_2(t)] \, [A_1^*(t+\tau) + A_2^*(t)] \rangle \tag{6.25}$$

where * denotes the complex conjugate. (To simplify this presentation complex amplitude transmission factors ('propagators') for the paths through the apertures, that take into account the size and distance of the apertures from P, have been omitted.)

The $\langle \rangle$ brackets in this equation are used to denote that measurement of intensity is (normally) a time-averaging process. The averaging is assumed to be over a time-span greater than the coherence time of the light.

Multiplying out gives

$$I_P = \langle A_1(t+\tau) A_1^*(t+\tau) \rangle + \langle A_2(t) A_2^*(t) \rangle$$
$$+ \langle A_1(t+\tau) A_2^*(t) \rangle \quad + \langle A_1^*(t+\tau) A_2(t) \rangle. \tag{6.26}$$

The first two terms are the intensities at C1 and C2 respectively, which can be denoted by I_P^{C1} and I_P^{C2}. (In the first term, τ simply shifts the time origin of the averaging and this normally has no effect on the value of the average: the system is assumed to be 'stationary'.)

The sum of the third and fourth terms is the sum of a complex number and its own complex conjugate, and is therefore just twice its real part. The equation can therefore be written as

$$I_P = I_P^{C1} + I_P^{C2} + 2\,\mathcal{R}(\Gamma_{12}(\tau)) \tag{6.27}$$

where $\mathcal{R}(\Gamma_{12}(\tau))$ is the real part of

$$\Gamma_{12}(\tau) = \langle A_1(t+\tau) A_2^*(t) \rangle \quad . \tag{6.28}$$

Partial Coherence, Correlation, and Visibility

The time-averaging in this definition of Γ_{12} can be expressed as an integration, and it is then obvious by reference to §4.7 that $\Gamma_{12}(\tau)$ can be described as the **complex cross-correlation function** between the illumination field at C1 and C2, in which the vibrations at C1 are 'considered' at time τ later than at C2. In the present context of physical optics $\Gamma_{12}(\tau)$ is also commonly referred to as the **complex mutual coherence function** of the illumination field at those points. $\mathcal{R}[\Gamma_{12}(\tau)]$ is then called the **mutual coherence**, and by comparison with our analysis of double-aperture diffraction in §1.1 its role in Eqn (6.27) clearly corresponds to what we recognized as the 'interference term' in Eqn. (1.01).

When C2 coincides with C1, Eqn (6.28) becomes

$$\Gamma_{11}(\tau) = \langle A_1(t+\tau) A_1^*(t) \rangle \tag{6.29}$$

which is the **complex self-coherence** (or **autocorrelation**) **function** of the field at C1. For $\tau = 0$, Γ_{11} reduces to just the intensity due to C1 alone, and in this simplified treatment we can write

$$I_P^{C1} = \Gamma_{11}(0)$$
$$I_P^{C2} = \Gamma_{22}(0) \quad . \tag{6.30}$$

Before proceeding, it is useful to normalize expressions such as Eqn (6.28), to enable coherence to be specified independently of the moduli of the amplitudes — they are really irrelevant to such a statement. The normalized complex coherence function is defined by the **complex *degree* of mutual coherence** (or **cross-correlation**), $\gamma_{12}(\tau)$, as follows

$$\gamma_{12}(\tau) = \frac{\langle A_1(t+\tau) A_2^*(t) \rangle}{[\langle |A_1|^2 \rangle \langle |A_2|^2 \rangle]^{1/2}} = \frac{\Gamma_{12}(\tau)}{\sqrt{\Gamma_{11}(0)} \sqrt{\Gamma_{22}(0)}} \quad . \tag{6.31}$$

It is also known as the **phase coherence factor**.

Using the Schwarz inequality in mathematics it can be shown that

$$0 \leqslant |\gamma_{12}(\tau)| \leqslant 1 \quad .$$

Eqn (6.27) then becomes

$$I_P = I_P^{C1} + I_P^{C2} + 2\sqrt{I_P^{C1}} \sqrt{I_P^{C2}} \, \mathcal{R}(\gamma_{12}(\tau)) \tag{6.32}$$

where $\mathcal{R}(\gamma_{12}(\tau))$ denotes the real part of $\gamma_{12}(\tau)$.

This equation is variously known as the *general interference law for stationary optical fields*, or as the *general interference law for partially coherent light*. An important feature is that it permits $\mathcal{R}(\gamma_{12}(\tau))$ to be calculated from experimentally measurable quantities. For any chosen pair of points in an illumination field a screen with suitably located pinholes is used just as in Fig. 6.07. The intensity, I_P, is measured at various positions of P to give an appropriate range of

τ, and the intensity at P from each pinhole alone, I_P^{C1} and I_P^{C2}, is also measured. The three quantities give $\mathcal{R}(\gamma_{12}(\tau))$ for each value of τ:

$$\mathcal{R}(\gamma_{12}(\tau)) = \frac{I_P - I_P^{C1} - I_P^{C2}}{2\sqrt{I_P^{C1}}\sqrt{I_P^{C2}}} \ . \tag{6.33}$$

The precise meaning of $\mathcal{R}(\gamma_{12})$ is made clearer by writing $\gamma_{12}(\tau)$ itself as

$$\gamma_{12}(\tau) = |\gamma_{12}(\tau)| e^{i\phi_{12}(\tau)} \tag{6.34}$$

where $\phi_{12}(\tau)$ refers to the generalized phase angle between the fields at P.

We can then write

$$\phi_{12}(\tau) = \alpha_{12}(\tau) + \delta \tag{6.35}$$

where δ is the phase angle associated solely with the internal path difference between C1 → P and C2 → P.

The real part of $\gamma_{12}(\tau)$ in Eqn (6.34) can now be expressed as follows

$$\mathcal{R}(\gamma_{12}(\tau)) = |\gamma_{12}(\tau)| \cos[\alpha_{12}(\tau) + \delta] \tag{6.36}$$

and Eqn (6.32) becomes

$$I_P = I_P^{C1} + I_P^{C2} + 2\sqrt{I_P^{C1}}\sqrt{I_P^{C2}} |\gamma_{12}(\tau)| \cos[\alpha_{12}(\tau) + \delta] \ . \tag{6.37}$$

The similarity to Eqn 1.01 is now even closer, and by comparing the two it is apparent that if $|\gamma_{12}(\tau)|$ has its upper extreme value, unity, then the intensity at P is the same as would be obtained with ideally monochromatic light and with a phase difference between the wavemotions at C1 and C2 equal to $\alpha_{12}(\tau)$. Under those circumstances the illumination field at C1 and C2 would be coherent. If $|\gamma_{12}(\tau)|$ has its other extreme value, zero, the final term in Eqn (6.37) is absent; there are no interference effects, because the illumination field at C1 and C2 is incoherent, and the observed intensity is simply the sum of the independent intensities due to C1 and C2. Most importantly, intermediate conditions, i.e. **partial coherence**, can be quantified by $|\gamma_{12}(\tau)|$, the **degree of mutual coherence**.

This treatment leads to a convenient concept of partial coherence as being equivalent to a mixture of coherent and incoherent light with intensities in the ratio

$$\frac{I_{\text{coh}}}{I_{\text{incoh}}} = \frac{|\gamma_{12}(\tau)|}{1 - |\gamma_{12}(\tau)|} \ . \tag{6.38}$$

However, the basically statistical nature of the physical phenomena that are being modelled in the whole of this treatment must be firmly remembered.

Returning to Eqn (6.37), we have not as yet seen how to deduce the modulus and argument of γ_{12} from experimental measurements; we have one equation and two unknowns. Let us reconsider our position. At the outset, and in order to obtain a general picture, the source was postulated as being both spatially and spectrally extended. All our considerations so far have respected that, and as a result the various equations referring to γ_{12} have been quite general with regard to the coherence of the illumination. Now Fig. 6.07 refers to a comparison between different points in the sampling plane of C1 and C2. This is clearly an arrangement that is especially sensitive to spatial (transverse) coherence. To relate γ_{12} to observables it is sensible, then, to consider the case where temporal coherence is not a complicating factor (§6.4.1). It is $\Gamma_{11}(\tau)$ that is appropriate for dealing specifically with temporal coherence since it is concerned with the extent to which phase relationships are maintained along individual wavetrains (§6.4.2).

6.4.1 Spatial coherence

Assume that the extended source in Fig. 6.07 is quasimonochromatic to the extent that the coherence time is much greater than the maximum value of τ associated with P for the range of positions of C1 and C2 over which we wish to assess the spatial coherence in the illumination field. Under those circumstances Eqn (6.37) gives maximum and minimum values of I_P when $\cos[\alpha_{12}(\tau)+\delta] = \pm 1$, since with the above assumption the background intensity level (and other factors dealt with in a rigorous analysis) will change only slowly through the fringe pattern. Furthermore, if τ itself is quite small it can be shown that the degree of mutual coherence, $|\gamma_{12}(\tau)|$, is inappreciably different from $|\gamma_{12}(0)|$, i.e. V and γ are largely independent of τ. Using all this gives the following expression for visibility

$$V_{12} = 2|\gamma_{12}| \frac{\sqrt{I_P^{C1}} \sqrt{I_P^{C2}}}{I_P^{C1} + I_P^{C2}} \ . \tag{6.39}$$

This is a relationship we have sought, and if the individual intensities associated with C1 and C2 are equal — as is often the case — then the *fringe visibility is equal to the degree of mutual coherence (cross-correlation)*, i.e.

$$V_{12} = |\gamma_{12}| \ . \tag{6.40}$$

Of Eqn (6.36) we need to note that the phase component α_{12} of the mutual coherence (i.e. any phase difference between the fields at C1 and C2) is revealed by the amount of off-axis shift in the position of the central fringe maximum, since the on-axis position corresponds to $\delta = 0$.

Experimental measurement of fringe visibility and fringe position therefore provides direct information about both the modulus and argument of the complex degree of spatial coherence in the illumination field given by an extended source.

With this direct relationship between visibility and correlation in mind we return to the similarity noted in §6.2.2 between two Fourier pairs, viz. the visibility–brightness distribution pair in Fig. 6.04 and the diffraction pattern–aperture function pair so familiar from previous chapters. As mentioned at the time, the similarity is not fortuitous or peculiar to that particular example. It can be shown that what one might call the 'pattern' of the complex degree of coherence (cross-correlation) in a plane illuminated by an extended source is the same as the complex-amplitude diffraction pattern given by an aperture of the same size and shape as that source. This is expressed formally in the **van Cittert–Zernike theorem,** to be found in the more advanced textbooks. Apart from its theoretical significance the theorem is particularly important from a computational point of view because the coherence calculation is normally more difficult than the calculation of the corresponding diffraction pattern.

Most importantly, a brightness distribution can be computed by Fourier transformation of the cross-correlation function derived from fringe visibility and phase data. And from our study of the spectral interferometer it is to be expected that a similar relationship exists between autocorrelation and spectral distribution. That is dealt with in the following section.

6.4.2 Temporal coherence

Here we show that the Fourier transform of the **complex self-coherence** (or **autocorrelation) function,** $\Gamma_{11}(\tau)$, is the spectral intensity distribution (the 'power spectrum'). We start by rewriting the time-averaging definition of $\Gamma_{11}(\tau)$ in Eqn (6.29) as an integration as follows

$$\Gamma_{11}(\tau) = \int A(t+\tau) \, A^*(t) \, dt \qquad (6.41)$$

where the earlier subscripts to A and A^* are implied. Then, by definition, the transform of $\Gamma_{11}(\tau)$ is

$$T[\Gamma_{11}(\tau)] = \int \Gamma_{11}(\tau) \, e^{2\pi i \nu \tau} \, d\tau \qquad (6.42)$$

where ν and τ are the frequency–time conjugate pair.

Substituting for Γ_{11} from Eqn (6.41) gives

$$T[\Gamma_{11}(\tau)] = \int_\tau \int_t A(t+\tau) \, A^*(t) \, e^{2\pi i \nu \tau} \, dt \, d\tau$$

$$= \int_t A^*(t) \, e^{2\pi i \nu t} \, dt \int_\tau A(t+\tau) \, e^{-2\pi i \nu (t+\tau)} \, d\tau \qquad (6.43)$$

$$= T[A^*(t)] \, T[A(t)] \, . \qquad (6.44)$$

(The second integral in Eqn (6.43) is with respect to τ and is independent of t

since the system is assumed to be stationary — cf. the discussion of Eqn (6.26).) Putting $F^*(\nu) = T[A^*(t)]$ and $F(\nu) = T[A(t)]$ we obtain

$$T[\Gamma_{11}(\tau)] = F^*(\nu) F(\nu)$$
$$= |F(\nu)|^2 \quad . \tag{6.45}$$

This equating of the Fourier transform of the autocorrelation function with the intensity (power) spectrum is an expression of the **Wiener–Khinchin theorem** (cf. §4.7.1).

The relationship parallels the result of the Michelson–Rayleigh treatment in §6.3.2. $|\Gamma_{11}(\tau)|$, or its normalised equivalent, $|\gamma_{11}(\tau)|$, can be related to visibility just as for $|\gamma_{12}|$ in Eqn (6.39); and the spectral intensity distribution, $|F(\nu)|^2$, was denoted in §6.3.2 by $f(\sigma')$. Visibility curves obtained with the two-beam spectral interferometer may be interpreted as representing $|\gamma_{11}(\tau)|$, a function of the path difference (corresponding to time τ) introduced in the comparision of a wavetrain with itself.

● ● ● ● ●

The elegance and 'symmetry' of the two pairs of Fourier transform relationships in interferometry are strikingly evident now. The visibility curve obtained in the spectroscopic application is in the time domain, i.e. it is a function of the time delay introduced between the two optical paths in the spectral interferometer in which a wavetrain is compared with itself (autocorrelation): the transform is the intensity ('power') spectrum of the source. With the stellar ('spatial') interferometer the visibility curve is a function of distance between two points in an illumination field that are compared (cross-correlation): its transform is the brightness spatial (angular) distribution of the source.

6.5 FOURIER TRANSFORM SPECTROSCOPY

In §6.3 we saw that with the Michelson spectral interferometer every wavetrain of light is amplitude-divided into two wavetrains that are then brought together after a path difference has been introduced between them. We saw that the way in which the visibility of the interference fringes changes as the path difference is altered (by translating mirror M_1 in Fig. 6.05) contains information about the spectral composition of the light. The routine practical application, however, had to await the arrival of the modern computer. Now, particularly because of the progress made in recent times in the development of small, dedicated computers and microprocessors as integral components of laboratory equipment, the advantages of 'Fourier spectroscopy' are being widely realized.

In this section we consider the Fourier aspect in more detail, in the context of some present-day techniques and applications.

The arrangement used in some types of modern Fourier transform spectrometer is shown in Fig. 6.08. It differs from that in Fig. 6.05 in one major respect; light from the source is collimated by a mirror C before being amplitude-divided by the beamsplitter B. This is the Twyman–Green version of the Michelson.

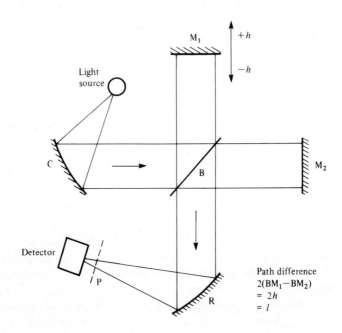

Fig. 6.08 – Fourier transform spectrometry.

The collimation has the effect of making the whole cross-section of the illumination field through the instrument correspond to the axial ($\theta = 0$) direction in Fig. 6.05. Circular fringes are therefore absent and the whole field is of uniform intensity. The changes in this intensity that occur as M_1 is translated are measured with the mirror and detector arrangement shown in the figure. Thus, in the case of the hypothetically monochromatic light that we considered, the detector would again record a sinusoidally varying intensity of illumination. If the wavenumber is σ_1 and the combining beams are of equal amplitude, A_1 say, then the intensity as a function of l is given by

$$I_{\sigma_1}(l) = 4 A_1^2 \cos^2(\pi \sigma_1 l)$$

corresponding to Eqn (6.17) with the path difference $l = 2h$, since $\theta = 0$. This can be rewritten as

$$I_{\sigma_1}(l) = 2 A_1^2 (1 + \cos 2\pi \sigma_1 l) \ . \tag{6.46}$$

The intensity measured by the detector for light of spectral intensity distribution $B(\sigma)$ is therefore given by

$$I(l) = \int_0^\infty B(\sigma)(1 + \cos 2\pi\sigma l)\, d\sigma \qquad (6.47)$$

omitting the numerical factor.

This integral can be written as

$$I(l) = \int_0^\infty B(\sigma)\, d\sigma + \int_0^\infty B(\sigma)\cos(2\pi\sigma l)\, d\sigma . \qquad (6.48)$$

When the two mirrors are exactly equidistant, optically, from the beamsplitter, $l = 0$ and $I(l) = I(0) = 2\int B(\sigma)\, d\sigma$. Therefore we have

$$I(l) - \tfrac{1}{2} I(0) = \int_0^\infty B(\sigma)\cos(2\pi\sigma l)\, d\sigma . \qquad (6.49)$$

This is a cosine Fourier transform relationship and we can write

$$B(\sigma) = \int_0^\infty [I(l) - \tfrac{1}{2} I(0)]\cos(2\pi\sigma l)\, dl . \qquad (6.50)$$

This equation, derived more rigorously in the specialist literature, is often described as the basic equation of Fourier transform spectroscopy. It enables the intensity spectrum, $B(\sigma)$, to be calculated for each chosen value of σ by performing the integration on the right.

Let us see what is involved. $I(l)$ is the intensity measured as a function of l. Plotted graphically it is called the 'interferogram', and Fig. 6.09(a) shows an example. (We shall discuss negative values of l presently.) Now in any interferogram, for $l = 0$ there is no path difference between combining wavefronts; amplitudes are added, and $I = I(0) = \sum (2A)^2 = 4\sum A^2$. At the other extreme, when l is so large that interference can no longer take place (i.e. l is greater than the coherence length of the light), then it is separate intensities that are added, giving $2\sum A^2 = \tfrac{1}{2} I(0)$. Inspection of Fig. 6.09(a) in the light of these identifications shows that $[I(l) - \tfrac{1}{2} I(0)]/\tfrac{1}{2} I(0)$ would be the 'visibility curve' (visibility as defined previously). We can therefore say that Eqn (6.50) relates $B(\sigma)$ to the Fourier transform of the visibility curve derived from the interferogram. Figure 6.09(b) shows the result for (a).

Note that in Eqn (6.50) the calculation of $B(\sigma)$ for each chosen wavenumber, σ_1 say, amounts to multiplying the whole of the curve $[I(l) - \tfrac{1}{2} I(0)]$ by $\cos(2\pi\sigma_1 l)$,

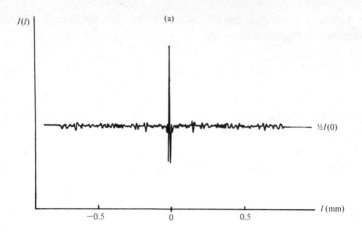

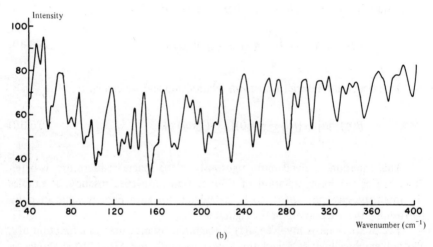

Fig. 6.09 – Interferogram (a) and calculated spectrum (b).
(Adapted from Bell (1972).)

the area under the product curve giving the required value of $B(\sigma_1)$. This is how the curve in Fig. 6.09(b) is obtained – though with a computer. It is reminiscent of the determination of Fourier coefficients in Chapter 3 (see Fig. 3.02) except that we are now dealing with non-repeating 'patterns' and therefore an integral rather than a series is involved. Also, we are working in the frequency–time domain, represented by σ and l respectively (l corresponds to a time delay).

Experimental measurements for negative values of l are as easy to make as for l positive, though the interferogram should be symmetrical. If, for any reason, it was not symmetrical, the cosine Fourier transform would not be applicable because it implicitly assumes that $I(l)$ is an even function. For reasons that are outside the scope of this book some instruments are in fact designed to be asymmetrical. In some, the specimen is placed in one arm of the instrument, such as in front of mirror M_2. Their use is concerned with 'amplitude spectroscopy', involving phase as well as amplitude, and requiring the calculation of the complex Fourier transform.

With regard to Eqn (6.50) it should be noted that it is assumed that the $l = 0$ position can be found accurately, and that the integration can be performed from $l = 0$ to $l = \infty$. The first of these is difficult and the latter impossible (Fig. 6.09). The more complicated methods mentioned above certainly help to overcome the first, and mathematical apodization is used as a means of minimizing the errors due to false termination of the Fourier transform integration (cf. the use of apodization in damping out the fringes around the Airy disc (§2.3)). However, the resolving power of the instrument, though limited, can clearly be very large since it is essentially determined by the upper limit of l available experimentally.

The interpretation of what is involved in this method can also be understood in terms of autocorrelation. We have just noted that the introduction of the path difference l amounts to a time delay — a time delay between the arrival of the two wavetrains that are brought together to interfere. This is the very essence of autocorrelation (§4.7), since each pair of wavetrains originates from a single wavetrain before it has encountered the beamsplitter; in effect, every initial wavetrain is compared with itself further along its length. The time delay, τ say, is given by l/c where c is the velocity of the light. The amplitudes of each pair of recombining wavetrains can therefore be written as $A(t)$ and $A(t + \tau)$. What is measured at each value of l then, is $\langle A(t)A(t + \tau)\rangle$. This is the autocorrelation function of the light, and we know from §6.4 that its Fourier transform is the power spectrum of the light, $|F(\nu)|^2$, (the Wiener–Khinchin theorem).

To return to the measurement of l. Alteration of l is achieved by the movement of a mirror, and the mechanical precision with which that can be controlled contributes towards setting a limit to the overall accuracy of the instrument. As a consequence, the method is particularly appropriate for work at long wavelengths, and it has been used extensively for some years in the infrared. However, with improvements in all aspects of the instrumentation the wavelength range has been extended down into the visible and even the ultraviolet.

Studies of the absorption by materials in the infrared are particularly important because of the type of information obtained. Materials absorb in that region for reasons associated with a number of phenomena. These include the vibrations of molecules (intra- and inter-molecular) and the rotations that occur about some chemical bonds.

Apart from their use in fundamental studies of materials, infrared spectra provide a powerful method of analysis. Figure 6.10 shows an example that is interesting from several points of view. In (b) we have the spectrum obtained

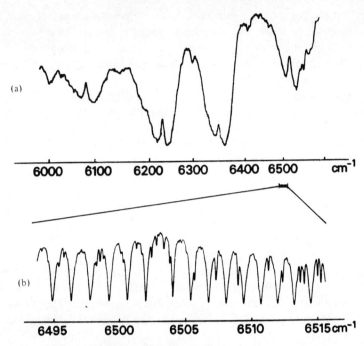

Fig. 6.10 – Spectra of light from Venus.
(a) Conventional grating spectroscopy.
(b) Fourier transform spectroscopy, showing detail of the rotational structure of the CO_2 band around 6505 cm^{-1}.
(After Connes, 1971)

by Fourier spectroscopy, of the light from Venus (and therefore something of a link with the next section). It reveals the absorption of light by carbon dioxide molecules in the atmosphere of the planet: all the details of the spectrum can be identified with the vibrational modes of CO_2. In (a) we have the spectrum obtained with a conventional grating-spectrometer.

Then again, the atomic groups attached to the carbon-atom skeletons of organic molecules are usually the sites of the chemical reactivity of the molecule. These 'functional groups' have characteristic vibration frequencies, and from an analytical point of view they provide a simple, rapid and reliable 'fingerprint' for assigning a compound to its chemical class.

The technique has complementary uses in inorganic chemistry, including studies of the solid-state physical properties of materials. It is also applicable to the investigation of plasma effects, relaxation phenomena, etc.

For absorption studies of materials in the laboratory the light-source is matched to the spectral region to be investigated, and the material whose absorption spectrum is required is placed in a specimen cell at P in Fig. 6.08. As a beam-splitter for the far-infrared (wavenumber below c. 400 cm^{-1} – see Appendix D) a sheet of Mylar or polyethylene may be used. For measurements between

200 cm^{-1} and 4000 cm^{-1} devices consisting of germanium films deposited on various substrates (e.g. KBr, CsI) have been developed. And in the region from 2 000 cm^{-1} to 16 000 cm^{-1} films of ferric oxide, Fe_2O_3, on a substrate such as CaF_2, are employed. A compensator plate (not shown) may be used.

The choice of detector again depends on the spectral region. A Golay cell, pyroelectric bolometer, or a detector of the semiconductor photoconducting type are typically used.

To minimize absorption by atmospheric water vapour the whole system is evacuated to 10^{-2} to 10^{-3} Torr. Even so, it is of course necessary to obtain interferograms with and without the specimen present, to correct for the characteristic interferogram of the instrument.

All spectroscopic work in the infrared has one feature in common: the low energy associated with the long wavelengths involved. In the far-infrared, where wavelengths extend into the radio microwave range, the energy of the radiation can be more than a thousand times lower than in the visible spectrum (cf. Appendix D). This is where, as we have seen, the interferometry method scores over the classical methods of spectroscopy which employ prism and grating spectrometers. There are two reasons for this. Firstly, as P. Jacquinot showed, all interferometers that have circular symmetry, such as the Michelson, accept a larger light flux than slit spectrometers of equal resolving power (Jacquinot and Dufour, 1948; see also Jacquinot, 1960). Secondly, the interferometry method 'multiplexes' a spectrum, by which is meant that it yields information about the whole of a spectrum simultaneously: this was put into quantitative terms by P. Fellgett in 1951 (see also Fellgett, 1958). During a traverse of the moveable mirror all the light entering the interferometer is used in the production of the interferogram and the latter contains all the information about the spectrum of the light. As a consequence, there is a considerable increase in the signal-to-noise ratio over that experienced with a dispersive instrument such as a grating spectrometer, where a spectrum is scanned wavelength by wavelength. (With some types of detector the signal-to-noise ratio can be further improved by signal-averaging techniques, and these are now commonly employed.)

In addition to spectroscopy in the infrared, and in the visible and ultraviolet, mentioned earlier, Fourier transform methods are now employed in other types of spectroscopy, including nuclear magnetic resonance spectroscopy (NMR), mass spectroscopy and a form of the latter known as ion cyclotron resonance spectroscopy (ICR).

6.6 APPLICATIONS IN ASTRONOMY

A glance at the expression

$$\theta = \frac{1.22\lambda}{a} \qquad \{2.07\}$$

for the angular resolution limit, θ, of a telescope of aperture diameter a, shows that in the radio spectrum of wavelengths in the range 10 cm–10^5 cm (Appendix D), a telescope reflector would need to have a diameter 2.10^5–2.10^9 times larger than that of the equivalent optical telescope working at a wavelength of about 500 nm. Added to that is the fact that whereas an optical telescope forms an image of stars in its field of view, a radio telescope gives only a single signal in the form of a voltage resulting from the radiation it receives over the entire range of directions determined by its resolving power. Small wonder that for some years after cosmic or galactic noise was discovered by K. G. Jansky in 1931 there was uncertainty as to where it was coming from.

Various observations were made by Jansky and others in the following years. By 1945 it was reported by E. V. Appleton that some radio emissions in the 60 MHz (5 m) band were associated with strong sunspot activity. In 1946 J. S. Hey and his co-workers discovered variations in the intensity of galactic noise from the direction of the constellation of Cygnus, suggesting that this particular radiation had its origins in a discrete source outside the solar system. So began the exploration of the radio sky.

This is certainly not the place to attempt to document the fascinating and spectacular history of radio astronomy. Nevertheless, notes on two of the important developments involving the use of Fourier methods are appropriate in this little book.

6.6.1 Aperture synthesis

In 1946 radar aerials had an angular beamwidth of about 10° at a wavelength of 1.5 m, which was quite inadequate for isolating, for example, regions on the surface of the sun from the general galactic background. Interferometry was a way out of this difficulty and J. L. Pawsey and his co-workers in Australia, making similar observations to Appleton's, used an aerial located high on a cliff overlooking the sea at Sydney. This arrangement (Fig. 6.11) formed an interferometer analogous to Lloyd's mirror experiment in optics. Interference occurred

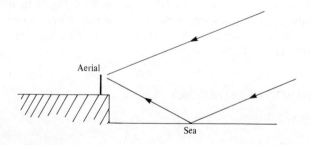

Fig. 6.11 – The sea interferometer.

[§6.6]   **Applications in Astronomy**   147

between direct radio signals and their reflection at the surface of the sea. The method was essentially the same as that involved in using a Michelson stellar interferometer but with the disadvantage of having a fixed baseline. Using the same aerial, J. Bolton and G. Stanley (1948) successfully recorded fringes produced by the Cygnus source — a constellation that conveniently rises only just above the horizon in Sydney. The workers in Australia also found other sources including a small intense source in the constellation of Taurus. This, alone among the early 'radio stars', was rapidly identified — with the Crab Nebula.

At the same time M. Ryle (now Sir Martin Ryle) and D. D. Vonberg, in Cambridge, were also studying radiation from the sun. They were using an interferometer in the form of a direct analogue of the Michelson (Fig. 6.12(a)). It had two aerial systems (dipole type) horizontally separated in the E–W direction, with the combined incoming signals fed to receiver equipment. Figure 6.12(b) shows the type of 'reception polar diagram' of such an interferometer, the envelope being the polar diagram of one aerial along. The reasoning behind this scheme was that if the separation between successive minima of this 'double-

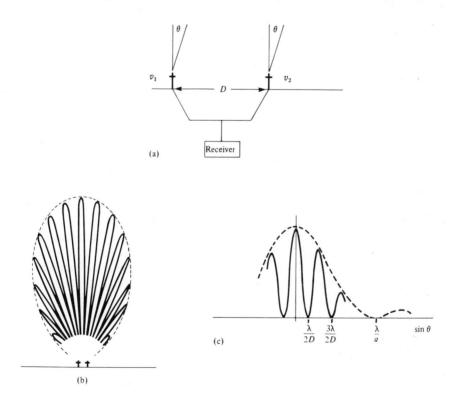

Fig. 6.12 – Radio analogue of Michelson stellar interferometer.

aperture' arrangement is made sufficiently large compared with the angular diameter of the sun, then as the aerial polar diagram is swept across the sun by the Earth's rotation any radiation from the sun would be detected by an oscillatory trace, whilst that from the galactic background would not.

The accuracy offered by this method is appreciated by considering Fig. 6.12(c) where we note that the angular width of the lobes in (b) is one half the width of the central maximum of a paraboloid aerial of diameter equal to the distance between the two aerials comprising the interferometer. To put it another way, the ability of the interferometer to locate sources and measure their angular diameters could only be matched by a conventional paraboloid radio telescope if it had a diameter equal to twice the separation of the two aerials used in the interferometer.

During the appearance of a large sunspot in 1946, when the solar radiation was considerably increased, Ryle and Vonberg used their apparatus to determine the angular diameter of the radio source in the sun. For different separations of the aerials they measured the ratio of the maximum to the minimum in the fringes comprising the oscillatory traces. From their results they deduced that the source had a diameter of 10 minutes of arc. As this value was not appreciably greater than the diameter of the visually seen sunspot they deduced that the radio source was related to the visual spot, or at least associated with it.

Using the two-aerial interferometer, Ryle and F. G. Smith (1948) discovered the strongest source in the northern sky, in Cassiopeia, but without any optical identification. They estimated that it had an angular diameter of less than 6 minutes of arc.

The Fourier aspect of variable-spacing interferometry for the derivation of the actual structure of a radio source was recognized by Pawsey and his collaborators in their work referred to above. It was used by H. M. Stanier in Cambridge in an investigation reported in 1950, to test theories developed in the late 1940s that emission from the sun in the absence of sunspots was particularly strong from the 'limb' of the sun at wavelengths of about 60 cm. In essentially the same way as that described in connection with the Michelson interferometer (§6.2.2) the fringe visibility was measured for aerial spacings up to 365 wavelengths. As the orientation of the aerial system was fixed the calculations had to be based on the assumption of circular symmetry in the source. Fourier transformation of the visibility curve gave a radial intensity distribution. (Strictly, this would be a Fourier–Bessel transformation.) Figure 6.13 shows the general form of the results, with no evidence of the limb-brightening that some investigators had expected.

The correlation aspect of the interferometer methods plays a very important role to which we now turn.

In Fig. 6.12(a) v_1 and v_2 are the voltages that are detected at the two aerials and fed to a receiver whose output is proportional to the average of their product. Now v_1 and v_2 each contain a 'noise' component because the individual aerials

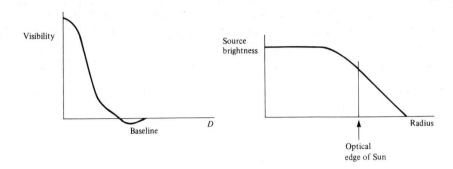

Fig. 6.13 — Visibility curve and radial distribution of radio 'brightness' across the solar disc.

(Based on Stanier, 1950)

have a large beamwidth. By its extraneous nature this noise is uncorrelated between 1 and 2. We can therefore write the time-average of the product of v_1 and v_2 as

$$\begin{aligned}\langle v_1 v_2 \rangle &= \langle (v_{u1} + v_{c1})(v_{u2} + v_{c2}) \rangle \\ &= \langle v_{u1} v_{u2} \rangle + \langle v_{u1} v_{c2} \rangle \\ &\quad + \langle v_{c1} v_{u2} \rangle + \langle v_{c1} v_{c2} \rangle \end{aligned} \quad (6.51)$$

where the subscripts u and c refer to uncorrelated and correlated signals respectively.

If the averaging time were sufficient all terms except the last one would be zero since they are products involving uncorrelated voltages. The time-averaged product of v_1 and v_2 would simply be

$$\langle v_1 v_2 \rangle = \langle v_{c1} v_{c2} \rangle. \quad (6.52)$$

However, the averaging time has to be finite, noise fluctuations do persist and the wanted signal is liable to be small compared with unwanted noise. To minimize this, Ryle (1952) introduced a phase-switching procedure in which the aerials were coupled alternately in and out of phase by means of a $\tfrac{1}{2}\lambda$ length of transmission line in the cable from one aerial (Fig. 6.12(a)). The difference between the two outputs was measured, giving

$$\langle (v_1 + v_2)^2 \rangle - \langle (v_1 - v_2)^2 \rangle = 4 \langle v_1 v_2 \rangle. \quad (6.53)$$

More elaborate methods have subsequently been devised but the principle remains the same.

Expressed in this way the interferometer is measuring the cross-correlation between the signals at the two aerials as a function of the aerial separation; its Fourier transform is the spatial brightness distribution of the source (§6.4.1).

A number of different **unfilled aperture** systems were developed for the study of the distribution of the galactic emission in order to improve angular resolution. It was only the ability to measure the cross-correlation (i.e. $\langle v_{c1} v_{c2} \rangle$) which allowed the polar-diagrams of two dissimilar aerials to be 'multiplied'. And it is *this* principle, whereby the effective area for 2-dimensional angular resolution no longer has to be the same as the area for sensitivity, that is implied by the term 'unfilled aperture'.

Ryle (1952) described the advantage of long arrays and various phase-switching schemes to improve performance. A notable example has been the 'Mills Cross' in Australia, named after its designer B. Y. Mills. Various forms of this interferometer were subsequently built in several countries. The original version (Mills and Little, 1953) consisted of two mutually perpendicular thin 'apertures' in the form of linear arrays running N–S and E–W (Fig. 6.14(a)). Each array comprised 250 dipoles (at 3.5 m wavelength) extending over a length of 1500 ft and consequently had a polar diagram rather like a fan, as shown in (b). Combined together, the region of interaction of the two fans formed a narrow 'pencil' beam and by an ingenious method of switching only the signals received in this narrow beam were measured. To alter the elevation of the beam, phase differences were introduced between all the elements in the N–S array by switching in the appropriate lengths of cable.

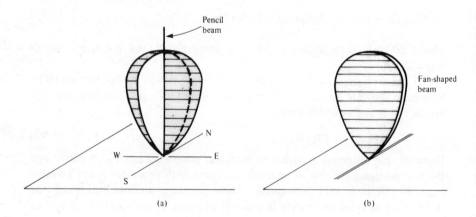

Fig. 6.14 – The Mills Cross.

So far we have outlined in very general terms some of the main optical aspects of achieving directional sensitivity at radio wavelengths, and how the 'unfilled aperture' methods succeed in that respect. But their collecting areas are relatively small and this makes them unsuitable for investigations requiring great sensitivity – where the critical factor is the smallest radio flux that can be detected with a system. In connection with the complex problem of making surveys of distant radio 'stars', in which the location of weak signals is involved, a team led by Ryle at the Mullard Radio Astronomy Observatory of the Cavendish Laboratory at Cambridge pioneered the use of the variable spacing interferometer for 'aperture synthesis'. The interferometer was used in such a way as to achieve not only the angular resolution of a single very large aerial but also to go some way to matching it in sensitivity too. The method, elaborated in a paper by Ryle and Hewish in 1960, can be visualized by thinking of the single large aperture that is ideally required, as divided into separate elements of area (Fig. 6.15).

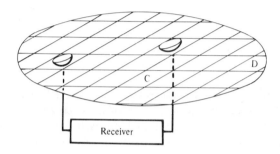

Fig. 6.15 — Aperture synthesis.

If the wavefronts received at these elements could be measured by placing a small radio telescope the size of just one of them, at each in turn; and if the signals could be combined in the correct phase; then the result would be that for the whole aperture. The phases between the separately measured signals could be determined by making the measurements in pairs, with the variable-spaced interferometer. As can be seen in the figure it is not necessary to make measurements for all the pairs; some (e.g. CD) are redundant. This type of synthesis was first accomplished in 1957 by Ryle's colleague J. H. Blythe, at a wavelength of 7.9 m and an effective angular resolution of approximately 1°.

Without the need for the two elements of the interferometer used in aperture synthesis to be equal in size and shape, various systems have evolved, for example employing a large fixed array in conjunction with a small, steerable parabolic aerial.

By utilizing the Earth's rotation a method of **'supersynthesis'** was initiated by Ryle in 1962. The principle is indicated in Fig. 6.16 where the two aerials A and B of an interferometer are separated by distance D along the E—W direction. The baseline rotates with the Earth every 24 hours and in one rotation each aerial has occupied all positions on an annular ring of radius D relative to the other. By using different values of D a complete circular area is synthesized, of radius equal to the maximum separation of the two aerials. (Strictly, the areas are ellipses except in the direction of the Pole star.)

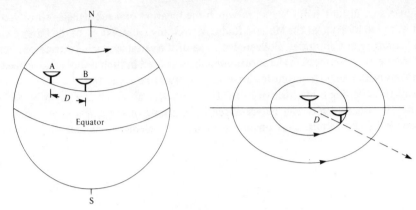

Fig. 6.16 – Supersynthesis.

In these synthesis methods the receiver circuitry, schematically shown in Fig. 6.17, allows the cosine and sine components, $C(D)$ and $S(D)$, of the interferometer receiver output to be measured for each value of D and projected angle of baseline. Reference to Eqns (6.12) and (6.13) will show that in the general case, and using exponential notation, we can write

$$C(D) + iS(D) = \int_\phi B(\phi) \, e^{2\pi i D \phi / \lambda} \, d\phi \qquad (6.54)$$

so that by Fourier transformation the brightness distribution $B(\phi)$ can be calculated. In practice it is of course theoretically and technically more complicated than this symbolic equation suggests. The problem is not only 3-dimensional but the aerials have a 'reception pattern' which modifies the $\cos^2$ response assumed in the derivation of the equations quoted above; and so on. The present description is intended only to convey the main principles.

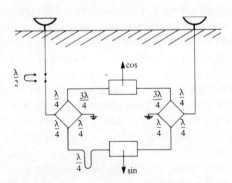

Fig. 6.17 – Sine and cosine components provide phase data.

The 5-km telescope at Cambridge has been used at a wavelength of 2 cm to give an angular resolution of 0.6 arc sec.

The Nobel Prize for physics was awarded in 1974 jointly to Ryle and Hewish for their pioneering research in radio astrophysics, Ryle for his observations and inventions, in particular the technique of aperture synthesis, and Hewish for the decisive role he played in the discovery of pulsars (— these were discovered with a different type of aerial, a massive array of 2048 dipoles covering an area of 18 km^2, that had been specially designed for the study of rapidly 'twinkling' radio sources for which aperture synthesis by its nature is unsuitable).

To observe radio galaxies and quasars in the finest detail aperture synthesis today uses baselines that extend across continents and oceans. Although radio links for bringing the signals together proved troublesome, because they posed problems of phase stability, the advent of accurate atomic clocks in the late 1960s has meant that the signals can be recorded separately on magnetic tape. The tape recordings can then be brought together and replayed to produce the required interferometer output. Even so, phase variations do occur and they have an effect that is equivalent to the positions of fringe maxima being fuzzy. Methods to overcome this difficulty have tended to be applicable only to the study of sources of very small size, or larger sources associated with a bright 'point' source, since this can act as a phase reference. The limitation imposed by this has now been overcome to a considerable extent by a new method. Dormant since the basis of it was realized in 1953 by R. C. Jennison at the Jodrell Bank Radio Observatory, it was conceived afresh and applied to long-baseline interferometry by A. E. Rogers in 1974 in the USA. This **phase closure** method is based on the fact that if phases are measured between pairs of aerials that form a closed loop, e.g. three in a triangle, then when they are added together the spurious effects cancel. With aerials located around the world the synthesis of a telescope the size of the Earth, and giving 'images' that are unaffected by atmospheric and other effects, has become a reality — with a resolution of 0.0001 seconds of arc obtainable. To put the latter in perspective one can note that it is approximately the angular diameter of the hot, visible stars; and with the best conventional optical telescopes their image is severely blurred by the effects of the turbulence in the Earth's atmosphere.

Fuller accounts of how these techniques are being used and details of the results that are being obtained (of which we have made no mention) are to be found in the literature cited in the Bibliography.

The correlation aspect of the interferometry methods led to another valuable development in astronomy, both radio and optical, to which we must now turn. It is one which appeared at first to contradict the requirements for observing interference effects: it combines intensities, not amplitudes.

6.6.2 The intensity interferometer

The principle of the intensity interferometer is due to R. Hanbury Brown who at Jodrell Bank in 1949 was considering two problems confronting radio astronomers: how to work with really long baselines, and how to overcome the effects of atmospheric turbulence. Mention of Jodrell Bank inevitably conjures up a picture of the massive 250 ft diameter steerable radio telescope. This was built in connection with a continuing programme of work by A. C. B. Lovell (now Sir Bernard Lovell) and his collaborators, on radar echoes from meteors and cosmic ray showers. In addition to that work this large dish-type aerial (of which there are now many of different shapes and sizes around the world) has played an important role in the development of various aspects of long-baseline interferometry.

In his book *The Intensity Interferometer* Hanbury Brown recalls how the challenge in 1949 was to measure the angular sizes of the two most prominent radio sources in the sky, Cygnus A and Cassiopeia A:

> At that time we knew only that their angular sizes could not be much greater than a few minutes of arc but we had no evidence as to how small they might be. If, as some people thought, they proved to be as small as the visible stars, then at metre wavelengths we should need two stations at the ends of the Earth. Was it feasible to build a radio interferometer with a baseline that could be extended if necessary from tens to hundreds, or perhaps thousands, of kilometres? The immediate technical difficulty in adapting conventional designs was to provide a coherent oscillator at the two distant points, and I started to wonder if this was really necessary. Could one perhaps compare the radio waves received at two points by some other means? As an example, I imagined a simple detector which demodulated waves from the source and displayed them as the usual noise which one sees on a cathode-ray oscilloscope. If one could take simultaneous photographs of the noise at two stations, would the two pictures look the same? This question led directly to the idea of the correlation of intensity fluctuations and to the principle of intensity interferometry.

The idea, therefore, was that if the fluctuations in intensity at two closely separated aerials correlated then the decrease in correlation (hence '**correlation interferometer**') as the baseline is increased would allow the angular diameter of the source to be determined (it would be the intensity analogue of Michelson's method for measuring the diameters of optically visible stars). The problem of the mutual instability of widely separated oscillators would be overcome. (At that time the atomic clocks now used in long-baseline interferometry had not been developed.)

The idea was successfully tried at Jodrell Bank (Hanbury Brown and Jennison, 1950), with an interferometer operating at a frequency of 125 MHz, to measure the angular diameter of the Sun. A year later it was used to measure the two radio sources that had posed the challenge. In the event they proved to be larger than expected and required only a baseline extension to a few kilometres. Though feeling that a sledgehammer had been used to crack a nut, Hanbury Brown and his collaborators found that when the radio sources were scintillating violently, due to ionospheric irregularities, the measurements of correlation were not significantly affected. The other problem had therefore been solved too.

Applications in Astronomy

The theory of the intensity radio-interferometer was put on a formal basis in 1954 by Hanbury Brown and R. Q. Twiss. Though of limited use in radio astronomy, because it is essentially a method in which the signal being measured must be strong compared with the noise level of the receivers, the important outcome has been in optical astronomy. It was the solution of the problem concerning atmospheric turbulence that prompted an enquiry into whether the intensity interferometer could usefully be employed in optical astronomy. Atmospheric turbulence[†], it will be recalled, was a serious handicap in Michelson's method for measuring stellar diameters.

In 1956 a version of the instrument (Fig. 6.18) using two searchlight mirrors (M_1 and M_2) of 1.56 m diameter and baseline variable up to 14 m was used to measure the angular diameter of Sirius for the first time. Individually the mirrors produced blurred images of 8 mm diameter which were directed on to the cathodes of photomultipliers (P_1 and P_2). After amplification the signals were multiplied together. The average value of the product over some hours gave a direct measure of the cross-correlation between the intensity fluctuations in the light at the two mirrors. The angular diameter of Sirius deduced from the loss of correlation with increasing baseline (cf. the loss of visibility in the Michelson method) was 0.0068 ± 0.0005 sec of arc, in good agreement with a value of 0.0063 ± 0.0006 sec of arc predicted by astrophysical theory on entirely different grounds.

The theory of the *optical* intensity-interferometer (Hanbury Brown and Twiss, 1958) is complicated by the quantum nature of the photoelectric effect.

In Fig. 6.18 let the instantaneous intensities at M_1 and M_2 be $I_1(t)$ and $I_2(t)$ respectively. The fluctuation in $I_1(t)$ can be expressed as $\Delta I_1(t) = I_1(t) - \langle I_1(t) \rangle$

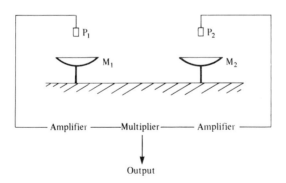

Fig. 6.18 – The intensity (or 'correlation') interferometer.

[†] For general-purpose optical astronomy the problems due to our atmosphere are substantially reduced in the specification of the Space Telescope. With a 2.4-m diameter mirror orbiting at an altitude of 500 km the maximum spatial resolution will be of the order of 0.1 arc sec, compared with ground-based telescopes which cannot better approximately 1 arc sec.

and similarly for I_2. Assuming the output currents from the photomultipliers to be proportional to the intensity of the light incident on them, the time-averaged product of the currents from the two photomultipliers is therefore proportional to $\langle \Delta I_1 \Delta I_2 \rangle$. Statistical calculations have shown that the latter is proportional to the square of the modulus of the cross-correlation between M_1 and M_2. Fourier transformation therefore gives the spatial brightness distribution of the source if, as is justifiable when measuring stellar diameters, it is assumed that the source is symmetrical.

Relatively insensitive to imperfections in mirrors and instability in the baseline, and less troubled than its forerunner with atmospheric disturbances, this has become an important technique in optical astronomy and has been actively pursued in more recent years by Hanbury Brown and his collaborators in Australia.

Appendices

APPENDIX A − THE SCALAR-WAVE DESCRIPTION OF ELECTRO-MAGNETIC WAVES

A.1 The scalar-wave description

In Fig. A.01(a), if P rotates about C with a uniform speed then its projection P' on Oy describes simple harmonic oscillations about O. If P' is attached to the end of a string, as illustrated in (b), the wavemotion transmitted along the string

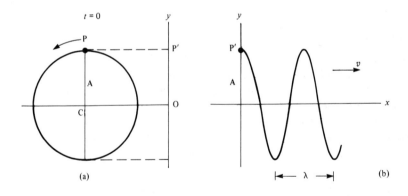

Fig. A.01

is analogous to the scalar-wave model for light (ignoring the question of polarization). To express this algebraically, we proceed as follows. First note that the oscillating displacement of P' about O is given by

$$\left.\begin{aligned} y &= A \cos \frac{2\pi t}{T} \\ &= A \cos 2\pi\nu t \\ &= A \cos \omega t \end{aligned}\right\} \qquad (A.01)$$

where A = amplitude of oscillation
T = period of oscillation (seconds)
ν = frequency (oscillations/sec)
ω = angular frequency (rads/sec) .

The equation for the wavemotion transmitted along the string may then be written in the form

$$y = A \cos\left[\omega t - \frac{2\pi x}{\lambda}\right] . \qquad (A.02)$$

The wave moves a distance of one wavelength (λ) for each oscillation and the velocity of the wave is therefore

$$v = \lambda/T$$

whence the usual relationship

$$v = \lambda\nu .$$

(The reader should confirm that the negative sign in Eqn (A.02) is associated with a wave travelling in the positive x direction, and vice versa.)

Eqn (A.02) is often written alternatively as

$$y = A \cos(\omega t - kx) \qquad (A.03)$$

where $k = 2\pi/\lambda$
and therefore $v = \omega/k$.

k is variously known as the angular wavenumber, wavelength constant or wave constant, and propagation constant. Often it is used to denote simply $1/\lambda$ and it is then called the (spectroscopic) wavenumber (the more conventional symbol, σ, for this is used in §6.3.2).

The wave considered so far has $y = A$ at $x = 0$ when $t = 0$. A similar wave is illustrated in Fig. A.02(b) but it is delayed relative to the one we have just considered by distance Δx. Its equation is therefore

$$y = A \cos\left[\omega t - \frac{2\pi}{\lambda}(x + \Delta x)\right] .$$

From Fig. A.02 it is evident that $\Delta x/\lambda = \alpha/2\pi$, and therefore

$$y = A \cos\left[\left(\omega t - \frac{2\pi x}{\lambda}\right) - \alpha\right]$$

$$= A \cos(\omega t - kx - \alpha) \qquad (A.04)$$

where $\alpha = \frac{2\pi}{\lambda}\Delta x \qquad (A.05)$

α is known as the **phase, phase angle**, or **phase constant**.

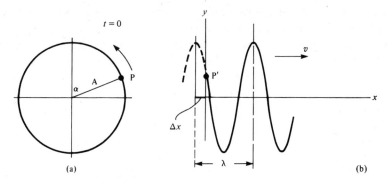

Fig. A.02

Relative to the wave represented by Eqn (A.03), α is referred to as the **phase difference** and Δx as the **path difference**. Thus we have the much used expression

$$\text{phase difference} = \frac{2\pi}{\lambda} \times \text{path difference}.$$

A.2 Combining waves
A.2.1 The principle of superposition: interference

Often in this book it is necessary to find the resultant illumination when a number of light waves, differing only in amplitude and phase, overlap at some point — on a screen, say. According to the principle of superposition the resultant 'disturbance' at that point is the sum of the separate disturbances.

Consider the two wavemotions

$$y_1 = A \cos(\omega t - kx)$$
$$y_2 = A \cos(\omega t - kx - \alpha)$$

where α is the phase difference between them.

Then

$$y_1 + y_2 = 2A \cos\frac{\alpha}{2} \cos\left(\omega t - kx - \frac{\alpha}{2}\right)$$

which is of the form

$$Y = R \cos(\omega t - kx - \theta) \quad . \tag{A.06}$$

This resultant is a light wave just like the separate incident waves, with frequency and velocity unaltered; but its amplitude, $2A\cos\alpha/2$, is not just the sum of the amplitudes of the separate incident waves. There is *interference* between the two, to an extent that depends on α. If the phase difference, α, between the two incident wavemotions is given by $\alpha = n2\pi$ (where n is zero or an integer), corresponding to zero path difference or a path difference of a whole number of wavelengths, the resultant does indeed have an amplitude of $2A$. However, if $\alpha = (n + \tfrac{1}{2})2\pi$ the two wavemotions are completely out of step and the resultant amplitude is zero. Phase differences between these two extremes give intermediate resultant amplitudes.

A similar result is obtained if the two incident amplitudes are unequal, though of course there could not be complete cancellation.

Note that if the above wavemotions separate again, beyond the place of intersection where they interfere, they proceed quite unaffected by the interference where they overlapped. Also note that *interference* can only occur between wavemotions having the same frequency.

For convenience, the place at which the effects of interference are to be calculated can be taken as $x = 0$. Eqn (A.04) becomes the displacement

$$y = A \cos(\omega t - \alpha) \quad . \tag{A.07}$$

The resultant displacement due to wavemotions of different amplitudes and phases can be expressed as

$$Y = \sum_n A_n \cos(\omega t - \alpha_n) \tag{A.08}$$

which takes the form

$$Y = R \cos(\omega t - \theta) \quad . \tag{A.09}$$

The summation is sometimes most easily performed with the aid of a phasor diagram as described in the following section, sometimes algebraically. For algebraic summation the use of exponential notation (§A2.3) is often simplest.

A.2.2 Phasor diagrams

Eqn (A.07) can be expanded as

$$y = A \cos\alpha \; \cos\omega t + A \sin\alpha \; \sin\omega t$$
$$= A_c \cos\omega t + A_s \sin\omega t$$

where $\quad A_c = A \cos\alpha, \quad A_s = A \sin\alpha \quad .$

A_c and A_s then carry the information about the amplitude and phase of the wave, as illustrated in Fig. A.03(a). We have

$$A = (A_c^2 + A_s^2)^{1/2}$$

and $\quad \alpha = \tan^{-1}(A_s/A_c) \quad .$

A_c and A_s are like the components of a two-dimensional vector but since angles here represent phase instead of direction the term *phasor* is used to make

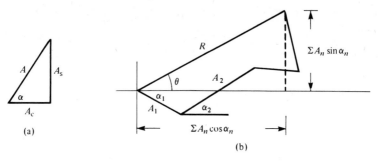

Fig. A.03

the distinction. The summation of Eqn (A.08) may therefore be performed graphically and Fig. A.03(b) shows that the resultant amplitude (R in Eqn (A.09)) is given by

$$R = \left[(\sum A_n \sin \alpha_n)^2 + (\sum A_n \cos \alpha_n)^2 \right]^{1/2}$$

and the phase $\quad \theta = \tan^{-1} \left[\dfrac{\sum A_n \sin \alpha_n}{\sum A_n \cos \alpha_n} \right].$ \hfill (A.10)

A.2.3 Exponential representation

A commonly used alternative procedure is based on expressing

$$y = A \cos(\omega t - \alpha)$$

in the form

$$y = \mathcal{R} A e^{i(\omega t - \alpha)} \tag{A.11}$$

where $\mathcal{R}$ indicates that only the real part of the expression is intended: for brevity $\mathcal{R}$ is usually omitted. $Ae^{-i\alpha}$ is called the **complex amplitude**.

The summation (Eqn (A.08))

$$Y = \sum_n A_n \cos(\omega t - \alpha_n)$$

is then written as

$$Y = \sum_n A_n e^{i(\omega t - \alpha_n)}$$

with $\mathcal{R}$ implied.

Appendix A

As before, the result of the summation is of the form

$$Y = R\cos(\omega t - \theta)$$

so that

$$Re^{i(\omega t-\theta)} = \sum_n A_n e^{i(\omega t - \alpha_n)}$$

i.e.
$$Re^{-i\theta} = \sum_n A_n e^{-i\alpha_n} \qquad (A.12)$$

$$= Z \text{ say}.$$

The complex amplitude, $Re^{-i\theta}$, of the sum is thus the sum of the complex amplitudes.

The summation is performed algebraically, and multiplying the expression that is obtained by its complex conjugate then gives R, since

$$R^2 = Z\,Z^* = |Z|^2.$$

This method is like the phasor method but with the summation performed in the complex plane. Thus, expanding Eqn (A.12)

$$R\cos\theta - iR\sin\theta = \sum A_n \cos\alpha_n - i\sum A_n \sin\alpha_n$$

and equating real and imaginary parts gives

$$R\cos\theta = \sum A_n \cos\alpha_n$$
$$R\sin\theta = \sum A_n \sin\alpha_n$$

whence
$$R = \left[\left(\sum A_n \cos\alpha_n\right)^2 + \left(\sum A_n \sin\alpha_n\right)^2\right]^{1/2}$$

$$\theta = \tan^{-1}\left[\frac{\sum A_n \sin\alpha_n}{\sum A_n \cos\alpha_n}\right]$$

as in Eqns (A.10).

Concerning the observation that a phasor diagram may be regarded as the addition of complex amplitudes in the complex plane, it should be noted that rotating a phasor through an angle α corresponds to multiplying the complex amplitude by $e^{i\alpha}$. It is because the latter type of operation is often mathematically easier than the manipulation of trigonometrical formulae that exponential notation is so often used.

APPENDIX B – THE STOKES TREATMENT OF REFLECTION AND REFRACTION

This classical treatment runs as follows. In Fig. B.01(a) a ray of light of amplitude A gives a reflection of amplitude Ar and a refracted ray of amplitude At at the interface between two dielectric media, 1 and 2, of different refractive indices. r and t denote the fraction of the amplitude reflected and refracted (transmitted) respectively.

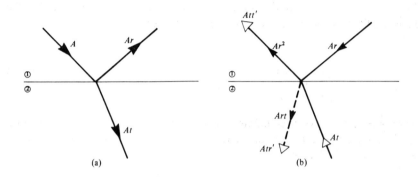

Fig. B.01

On the basis of the reversibility of light rays, if these reflected and refracted rays are reversed they should amount to a reversal of the incident ray of amplitude A. This is shown schematically in (b). The reversal of the ray of amplitude Ar gives a reflection back into the first medium of amplitude Ar^2, and a refracted ray of amplitude Art. Similarly, the reversal of the ray of amplitude At gives a reflection back into the second medium of amplitude Atr' and a transmitted ray into the first medium of amplitude Att', where r' and t' are fractions appropriate for events at the 2 / 1 interface as opposed to the 1 / 2 interface.

To reproduce (a) in reverse we therefore have to equate as follows

$$Att' + Ar^2 = A$$

i.e.
$$tt' = 1 - r^2$$

and
$$Art + Atr' = 0$$

i.e.
$$r = -r' \; .$$

The difference in sign of the amplitudes in this last equation indicates a phase change of π in the reflection at either 1 / 2 or 2 / 1. Experimental observation shows that it normally occurs when light is incident at a boundary from the side of higher velocity (lower refractive index).

APPENDIX C – DIFFRACTION OF X-RAYS BY CRYSTALS. THE EQUIVALENCE OF THE LAUE CONDITIONS AND THE BRAGG REFLECTION CONCEPT

The rectangular parallelepiped in Fig. C.01 depicts the unit cell of a crystal lattice, with each lattice 'point' representing an identical group of atoms. In §2.7 it was noted that an X-ray diffraction maximum occurs in a given direction from a crystal with this lattice if the X-ray scattering in that direction from the atomic group associated with each of the lattice points A, B and C is in phase with that scattered from the group associated with the lattice point O. The Laue conditions (Eqn (2.18)) are a statement of this requirement, viz.

$$\left. \begin{array}{l} (s - s_0) \cdot a = h\lambda \\ (s - s_0) \cdot b = k\lambda \\ (s - s_0) \cdot c = l\lambda \end{array} \right\} \quad (C.01)$$

where s_0, s are unit vectors defining the incident and scattered X-ray directions; a, b, c are the lattice translation vectors; h, k, l are integers or zero; and λ is the wavelength of the X-rays.

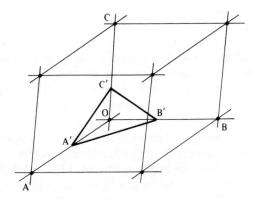

Fig. C.01

In explaining his 'reflection' description of the same phenomenon, W. L. Bragg pointed out that the requirement that scattering from just A, B and C be in phase is analagous to optical reflection in a mirror whose surface contains A, B, C: for any angle of incidence, reflection occurs at an angle equal to the angle of incidence (Fig. C.02). As Bragg explained, this relation of reflected to incident wave ensures that the waves scattered by all points in two directions in space, over the lattice plane, are in phase with each other. Unlike optical reflection, however, when X-rays are incident at a crystal lattice plane only a very small fraction of the amplitude of the incident beam is used in this way. Most of the

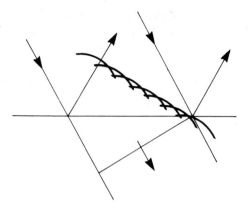

Fig. C.02 – The reflection of a wavefront by scattering centres lying in a plane (shown edge-on).

radiation is transmitted through the crystal. Furthermore, reflections from the successive lattice planes parallel to the first will not, in general, be in phase with each other. Reinforcement can be obtained, however, by adjusting the angle of incidence. As shown in Fig. C.03 it requires that the path difference X′Y′Z–XYZ is a whole number of wavelengths. Since Y′Y = Y′W, this is equivalent to requiring that

$$VW = n\lambda$$

i.e. $\qquad 2d \sin \theta = n\lambda \qquad$ (C.02)

where d is the spacing of the lattice planes concerned, and the integer n is the 'order' of the reflection.

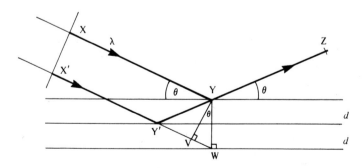

Fig. C.03 – The reflection of a wavetrain by successive lattice-planes of spacing d.

Appendix C

This is the familiar 'Bragg equation' and the special values of the glancing angle θ are the 'Bragg angles'. When the condition expressed in the equation is fulfilled the scattered X-ray waves from all lattice points throughout the crystal reinforce each other: at other angles of incidence there is destructuve interference. As Bragg pointed out, in his penetratingly clear way, the equation is 'the familiar optical relation giving the colours reflected by thin films, in another guise' (Bragg, 1975).

To establish the equivalence of the Laue and Bragg interpretations, and to find which are the reflecting planes corresponding to particular values of h, k, l in Eqn C.01, we proceed as follows.

In Fig. C.01 construct the plane $A'B'C'$, such that $OA' = OA/h$, $OB' = OB/k$, $OC' = OC/l$, where h, k, l are the integers in the Laue equations. Since h, k, l are integers it follows that A', B', C' lie in a lattice plane that passes through lattice points elsewhere in the same lattice (of which we have only drawn one unit cell). And since in the derivation of the Laue equations it was stated that, for a diffraction maximum to occur, the scattering from A, B, C would be h, k, l wavelengths ahead (or behind) that from O, it follows that the scattering from A', B', C' would be just one wavelength different.

A more convenient view of $A'B'C'$ for the next step is shown in Fig. C.04(a) where part of the next lattice-plane parallel to $A'B'C'$, the one passing through O, is also shown. We are now ready to look again at the Laue equations.

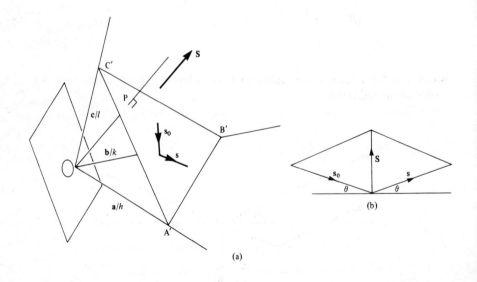

Fig. C.04

Putting $s - s_0 = S$, Eqns (C.01) become

$$\left. \begin{array}{c} \dfrac{S}{\lambda} \cdot \dfrac{a}{h} = 1 \\[1em] \dfrac{S}{\lambda} \cdot \dfrac{b}{k} = 1 \\[1em] \dfrac{S}{\lambda} \cdot \dfrac{c}{l} = 1 \end{array} \right\} \qquad (\text{C.03})$$

Subtracting the second from the first gives

$$\frac{S}{\lambda} \cdot \left(\frac{a}{h} - \frac{b}{k} \right) = 0$$

i.e. **S** is perpendicular to $\overrightarrow{B'A'}$.

Subtracting the other pairs in Eqn (C.03) shows that **S** is also perpendicular to $\overrightarrow{A'C'}$ and $\overrightarrow{C'B'}$. **S** is therefore perpendicular to the plane $A'B'C'$. Now in Fig. C.04(b) we see that **S** bisects **s** and s_0. It follows, therefore, that **s** and s_0 are equally inclined to $A'B'C'$, in accord with the idea of reflection in the lattice plane $A'B'C'$.

As points in the plane $A'B'C'$ scatter with a path difference of one wavelength compared with those in the parallel lattice-plane through O, we also have Bragg's picture of the reinforcement of scattering from successive lattice planes. If s_0 and **s** are at the equal glancing angle to $A'B'C'$ that Bragg designates as θ, we should therefore be able to show that the spacing, OP, of these planes is equal to the d in the Bragg equation, for $n = 1$. From the figure we have

$$OP = \frac{a}{h} \cdot \frac{S}{|S|} .$$

With the first of the Laue equations this gives

$$OP = \frac{\lambda}{|S|} .$$

From Fig. C.04(b) we have $|S| = 2 \sin\theta$ and therefore

$$OP = \frac{\lambda}{2 \sin\theta}$$

$\qquad = d$ in the Bragg equation, for $n = 1$.

Thus we have shown that (i) scattered X-rays comprising a diffraction maximum according to the Laue equations also constitute a reflection in the optical sense, the reflection being at lattice planes defined by the h, k, l values in the Laue equations, and (ii) the spacing of those lattice planes, and the angle of reflection at them, are as given by the Bragg equation.

APPENDIX D – THE ELECTROMAGNETIC SPECTRUM.

APPROXIMATE RANGES OF PRINCIPAL, NAMED REGIONS

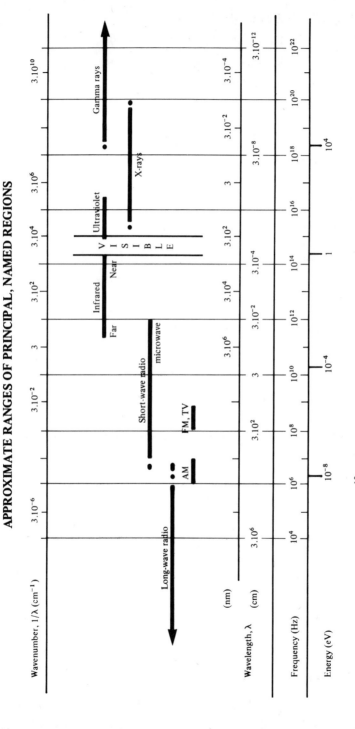

Velocity taken as approximately 3.10^{10} cm/s.
1 eV has associated frequency c. 2.42×10^{14} Hz, wavelength c. 1240 nm.

APPENDIX E – USEFUL FORMULAE

$$\sin(A \pm B) = \sin A \cos B \pm \cos A \sin B$$

$$\cos(A \pm B) = \cos A \cos B \mp \sin A \sin B$$

$$\sin 2\theta = 2 \sin \theta \cos \theta$$

$$\cos 2\theta = \cos^2 \theta - \sin^2 \theta$$

$$= 2 \cos^2 \theta - 1$$

$$= 1 - 2 \sin^2 \theta$$

$$\sin A + \sin B = 2 \sin \tfrac{1}{2}(A + B) \cos \tfrac{1}{2}(A - B)$$

$$\sin A - \sin B = 2 \cos \tfrac{1}{2}(A + B) \sin \tfrac{1}{2}(A - B)$$

$$\cos A + \cos B = 2 \cos \tfrac{1}{2}(A + B) \cos \tfrac{1}{2}(A - B)$$

$$\cos A - \cos B = -2 \sin \tfrac{1}{2}(A + B) \sin \tfrac{1}{2}(A - B)$$

$$e^{i\theta} = \cos \theta + i \sin \theta \text{ (Euler's relation)}^\dagger$$

$$\cos \theta = \frac{e^{i\theta} + e^{-i\theta}}{2}$$

$$\sin \theta = \frac{e^{i\theta} - e^{-i\theta}}{2i}$$

---

†Described by Richard Feynman as 'this amazing jewel ... the most remarkable formula in mathematics'.

References

Abbe, E. (1873) Beiträge zur theorie des mikroskops und der mikroskopischen wahrnehmung. *Archiv. f. Microskop. Anat.*, **9**, 413–468. (*Collected Works* (1904) Vol. 1.)

Barer, R. (1955) Phase contrast microscopy. *Research*, **8**, 341–343.

Blythe, J. H. (1957) A new type of pencil beam aerial for radio astronomy. *Roy. Ast. Soc. Monthly Notices*, **117**, 644–651.

Bolton, J. G. and Stanley, G. J. (1948) Variable source of radio frequency radiation in the constellation of Cygnus. *Nature*, **161**, 312–313.

Bragg, W. L. (1929) The determination of parameters in crystal structures by means of Fourier analysis. *Proc. Roy. Soc. A*, **123**, 537–559.

Bragg, W. L. (1929) An optical method of representing the results of X-ray analysis. *Z. f. Krist.* **70**, 475–492.

Bragg, W. L. (1933) *The crystalline state*, Vol. I. Bell, London.

Bragg, W. L. (1939) A new type of 'X-ray microscope'. *Nature*, **143**, 678.

Bragg, W. L. (1942) The X-ray microscope. *Nature*, **149**, 470–471.

Bragg, W. L. (1944) Lightning calculations with light. *Nature*, **154**, 69–72.

Buerger, M. J. (1950) Generalised microscopy and the two-wavelength microscope. *J. Appl. Phys.*, **21**, 909–917.

Connes, P. (1971) Advances in Fourier spectroscopy. *Phys. Bull.*, **22**, 26–28.

DeRosier, D. J. and Klug, A. (1968) Reconstruction of three-dimensional structures from electron micrographs. *Nature*, **217**, 130–134.

DeRosier, D. J. and Klug, A. (1972) Structure of the tubular variants of the head of bacteriophage T4 (polyheads). *J. Mol. Biol.*, **65**, 469–488.

Elias, P. (1953) Optics and communication theory. *Jour. Opt. Soc. Amer.*, **43**, 229–232.

Elias, P., Grey, D. S. and Robinson, D. Z. (1952) Fourier treatment of optical processes. *Jour. Opt. Soc. Amer.*, **42**, 127–134.

Fellgett, P. (1951) PhD thesis, University of Cambridge, UK.

Fellgett, P. (1958) A propos de la théorie du spectromètre interférentiel multiplex. *J. Phys. Radium*, **19**, 187.

References

Fizeau, H. (1862) *Ann. Chim. Phys.,* **66**, 429.

Fizeau, H. (1868) *C. R. Acad. Sci. Paris,* **66**, 934.

Gabor, D. (1948) A new microscope principle. *Nature,* **161**, 777–778.

Hanbury Brown, R. and Jennison, R. C. (1950) Unpublished.

Hanbury Brown, R. and Twiss, R. Q. (1954) A new type of interferometer for use in radio astronomy. *Phil. Mag.,* **45**, 663–682.

Hanbury Brown, R. and Twiss, R. Q. (1956) A test of a new type of stellar interferometer on Sirius. *Nature,* **178**, 1046–1048.

Hanbury Brown, R. and Twiss, R. Q. (1958) Interferometry of the intensity fluctuations in light – III. Applications to astronomy. *Proc. Roy. Soc. A,* **248**, 199–221.

Hanson, A. W. and Lipson, H. (1952) A simplified fly's eye procedure. *Acta Cryst.,* **5**, 145.

Harburn, G., Taylor, C. A. and Welberry, T. R. (1975) *Atlas of optical transforms.* Bell, London.

Hopkins, H. H. (1951) The concept of partial coherence in optics. *Proc. Roy. Soc. A,* **208**, 263–277.

Jacquinot, P. (1960) New developments in interference spectroscopy. *Rep. Progr. Phys.,* **23**, 267–312.

Jacquinot, P. and Dufour, C. J. (1948) Conditions optiques d'emploi des cellules photo-électriques dans les spectrographes et les interféromètres. *J. Rech. C.N.R.S.,* **6**, 91–103.

Klug, A. and Berger, J. E. (1964) An optical method for the analysis of periodicities in electron micrographs, and some observations on the mechanism of negative staining. *J. Mol. Biol.,* **10**, 565–569.

Klug, A. and Finch, J. T. (1968) Structure of viruses of the papilloma-polyoma type. *J. Mol. Biol.,* **31**, 1–12.

Leith, E. N. and Upatnieks, J. (1962) Reconstructed wavefronts and communication theory. *J. Opt. Soc. Amer.,* **52**, 1123–1130.

Leith, E. N. and Upatnieks, J. (1963) Wavefront reconstruction with continuous-tone objects. *J. Opt. Soc. Amer.,* **53**, 1377–1381.

Leith, E. N. and Upatnieks, J. (1964) Wavefront reconstruction with diffused illumination and 3-dimensional objects. *J. Opt. Soc. Amer.,* **54**, 1295–1301.

Leith, E. N. and Upatnieks, J. (1965) Wavefront reconstruction photography. *Physics Today,* **18**, No. 8, 26–32.

Michelson, A. A. (1890a) On the application of interference methods to astronomical measurements. *Phil. Mag.,* Ser 5, **30**, 1–21.

Michelson, A. A. (1890b) Measurement by light-waves. *Am. Jnl. Sci.,* **39**, 115–121.

Michelson, A. A. (1891a) Visibility of interference-fringes in the focus of a telescope. *Phil. Mag.,* Ser. 5, **31**, 256–259.

Michelson, A. A. (1891b) On the application of interference-methods to spectroscopic measurements – I. *Phil. Mag.,* Ser 5, **31**, 338–346.

References

Michelson, A. A. (1892) On the application of interference-methods to spectroscopic measurements – II. *Phil. Mag.*, Ser 5, **34**, 280–299.

Michelson, A. A. (1920) On the application of interference methods to astronomical measurements. *Astrophys. Jnl*, **51**, 257–262.

Michelson, A. A. (1927) *Studies in optics.* University of Chicago Press, Chicago. (Phoenix Science Series edn. 1962.)

Mills, B. Y. and Little, A. G. (1953) A high resolution aerial system of a new type. *Australian J. Phys.*, **6**, 272–278.

Padawer, J. (1968) The Nomarksi interference-contrast microscope. An experimental basis for image interpretation. *J. Roy. Mic. Soc.*, **88**, 305–349.

Pawsey, J. L., Payne-Scott, R. and McCready, L. L., (1946) Radio-frequency energy from the Sun. *Nature*, **157**, 158–159.

Porter, A. B. (1906) On the diffraction theory of microscopic vision. *Phil. Mag.*, Ser 6, **11**, 154–166.

Pykett, I. L. (1982) NMR imaging in medicine. *Sci. Amer.*, **246**, 54–64.

Rayleigh, Lord (1874) On the manufacture and theory of diffraction-gratings. *Phil. Mag.*, **47**, 81–93, 193–205. (*Scientific Papers*, Vol. I, 199–221. CUP, Cambridge.)

Rayleigh, Lord (1892) On the interference bands of approximately homogeneous light: in a letter to Prof. A. Michelson. *Phil. Mag.* Ser 5, **34**, 407–411. (*Scientific Papers*, Vol. IV, 15–18. CUP, Cambridge.)

Rayleigh, Lord (1896) On the theory of optical images, with special reference to the microscope. *Phil. Mag.*, **42**, 167–195. (*Scientific Papers*, Vol. IV, 235–260. CUP, Cambridge.)

Rayleigh, Lord (1903) On the theory of optical images, with special reference to the microscope. *Journ. R. Micr. Soc.*, 474–482. (*Scientific Papers*, Vol. V, 118–125. CUP, Cambridge.)

Ryle, M. (1952) A new radio interferometer and its application to the observation of weak radio stars. *Proc. Roy. Soc. A*, **211**, 351–375.

Ryle, M. and Vonberg, D. D. (1946) Solar radiation on 175 Mc/s. *Nature*, **158**, 339–340.

Ryle, M. and Smith, F. G. (1948) A new intense source of radio-frequency radiation in the constellation of Cassiopeia. *Nature*, **162**, 462–463.

Ryle, M. and Hewish, A. (1960) The synthesis of large radio telescopes. *Roy. Ast. Soc. Monthly Notices*, **120**, 220–230.

Schade, O. H. (1948) Electro-optical characteristics of television systems. *RCA Review*, **9**, 5–37, 245–286, 490–530, 653–686.

Shannon, C. (1949) *The mathematical theory of communication.* University of Illinois, Urbana.

Stanier, H. M. (1950) Distribution of radiation from the undisturbed Sun at a wave-length of 60 cm. *Nature*, **165**, 354–355.

Taylor, C. A. and Lipson, H. (1964) *Optical transforms.* Bell, London.

Vander Lugt, A. B. (1963) Signal detection by complex spatial filtering. *Radar Lab., Report* No. 4594–22–T, University of Michigan, Ann Arbor.

Vander Lugt, A. B. (1964) Signal detection by complex spatial filtering. *IEEE Trans. Inform. Theory,* **IT–10**:2.

Wiener, N. (1949) *The extrapolation, interpolation and smoothing of stationary time series.* Technology Press and Wiley, New York.

Zernike, F. (1938) The concept of degree of coherence and its applications to optical problems. *Physica,* **5**, 785–795.

Zernike, F. (1942) Phase contrast, a new method for the microscopic observation of transparent objects. *Physica,* **9**, 686–698, 974–986.

Bibliography

Bahcall, J. N. and Spitzer, L. (1982) The space telescope. *Sci. Amer.*, **247**, 38–49.
Barnes, K. R. (1971) *The optical transfer function*. Hilger, London.
Bell, R. J. (1972) *Introduction to Fourier transform spectroscopy*. Academic Press, New York.
Born, M. and Wolf, E. (1980) *Principles of optics*, 6th edn. Pergamon, Oxford.
Bracewell, R. N. (1979) *The Fourier transform and its applications*, 2nd edn. McGraw-Hill, New York.
Bragg, W. L. (Phillips, D. C. and Lipson, H. (eds.)) (1975) *The development of X-ray analysis*. Bell, London.
Brown, E. B. (1965) *Modern optics*. Reinhold, New York.
Casasent, D. (ed.) (1978) *Optical data processing*. Springer, Berlin.
Christiansen, W. N. and Högbom, J. A. (1969) *Radiotelescopes*. CUP, London.
Clarke, R. H. and Brown, J. (1980) *Diffraction theory and antennas*. Ellis Horwood, Chichester, UK.
Culling, C. F. A. (1974) *Modern microscopy: elementary theory and practice*. Butterworth, London.
Cutrona, L. J., Leith, E. N., Palermo, C. J. and Porcello, L. J. (1960) Optical data processing and filtering systems. *IRE Trans. Inform. Theory*, **IT-6**, 386–400.
Ditchburn, R. W. (1976) *Light*, 3rd edn. Blackie, London.
Duffieux, P. M. (1983) *The Fourier transform and its applications to optics*, 2nd edn. Wiley, New York.
Duffieux, P. M. (1946) *L'intégrale de Fourier et ses applications à l'optique*. Published privately: Rennes, France.
Fowles, G. R. (1975) *Introduction to modern optics*, 2nd edn. Holt, Rinehart and Winston, New York.
Françon, M. (1963) *Modern applications of physical optics*, English edn. (transl. Scripta Technica). Wiley, New York.
Françon, M. (1966) *Diffraction: coherence in optics*, English edn. (transl. Jeffrey, B. and edited Sanders, J. H.). Pergamon, London.

Françon, M. (1966) *Optical interferometry,* English edn. (transl. Wilmanns, I.). Academic Press, New York.
Françon, M. (1974) *Holography,* English edn. (transl. Spruch, G. M.). Academic Press, New York.
Françon, M. (1979) *Optical image formation and processing,* English edn. (transl. Jaffe, B. M.). Academic Press, New York.
Gabor, D. (1972) (*Edwards Memorial Lecture*) *Holography: 1948–1971.* The City University, London.
Gaskill, J. D. (1978) *Linear systems, Fourier transforms and optics.* Wiley, New York.
Ghatak, A. (1977) *Optics.* Tata McGraw-Hill, New Delhi.
Goodman, J. W. (1968) *Introduction to Fourier optics.* McGraw-Hill, New York.
Graham, N. (1979) Does the brain perform a Fourier analysis of the visual scene? *TINS,* 2, 207–208.
Griffiths, P. R. (ed.) (1978) *Transform techniques in chemistry.* Heyden, London.
Hanbury Brown, R. (1974) *The intensity interferomter.* Taylor and Francis, London.
Harburn, G., Taylor, C. A. and Welberry, T. R. (1975) *Atlas of optical transforms.* Bell, London.
Hawkes, P. W. (ed.) (1980) *Computer processing of electron microscope images.* Springer, Berlin.
Hecht, E. and Zajac, A. (1974) *Optics.* Addison-Wesley, Reading, Mass.
Hewish, A. (1965) The synthesis of giant radio telescopes. *Sci. Progr.,* 53, 355–368.
Hey, J. S. (1971) *The radio universe.* Pergamon, Oxford.
Hey, J. S. (1973) *The evolution of radio astronomy.* Elek Science, London.
Hopkins, H. H. (1962) 21st Thomas Young Oration. The application of frequency response techniques in optics. *Proc. Phys. Soc.,* 79, 889–919.
Jenkins, F. A. and White, H. E. (1976) *Fundamentals of optics.* 4th edn. McGraw-Hill Kogakusha, Tokyo.
Jennison, R. C. (1966) *Introduction to radio astronomy.* Newnes, London.
Kihlborg, L. (ed.) (1979) *Nobel Symposium 47. Direct imaging of atoms in crystals and molecules.* The Royal Swedish Academy of Sciences, Stockholm.
Klein, M. W. (1970) *Optics.* Wiley, New York.
Kraus, J. D. (1966) *Radio astronomy.* McGraw-Hill, New York.
Lee, S. H. (ed.) (1981) *Optical information processing – fundamentals.* Springer, Berlin.
Lipson, H. S. (1970) *Crystals and X-rays.* Wykeham Publications, London.
Lipson, H. (ed.) (1972) *Optical transforms.* Academic Press, London.
Lipson, H. and Taylor, C. A. (1958) *Fourier transforms and X-ray diffraction.* Bell, London.
Lipson, S. G. and Lipson, H. (1981) *Optical physics,* 2nd edn. CUP, London.

Longhurst, R. S. (1973) *Geometrical and physical optics.* 3rd edn. Longman, London.
McLachlan, D. (1962) The role of optics in applying correlation functions to pattern recognition. *J. Opt. Soc. Amer.,* **52**, 454–459.
Marathay, A. S. (1982) *Elements of optical coherence theory.* Wiley, New York.
Michelson, A. A. (1927) *Studies in optics.* University of Chicago Press, Chicago. (Phoenix Science Series edn, 1962.)
Mertz, L. (1965) *Transformations in optics.* Wiley, New York.
O'Neill, E. L. (1956) Spatial filtering in optics. *IRE Trans. Inform. Theory,* **IT–2**, 56–65.
O'Neill, E. L. (1963) *Introduction to statistical optics.* Addison-Wesley, Reading, Mass.
Readhead, A. C. S. (1982) Radio astronomy by very-long-baseline interferometry. *Sci. Amer.,* **247**, 39–47.
Rogers, G. L. (1954) The Abbe theory of microscopic vision and the Gibbs phenomenon. *Amer. Jnl. Phys.,* **22**, 384–389.
Schawlow, A. L. (ed.) (1969) *Lasers and light* (Readings from *Scientific American*). Freeman, San Francisco.
Shannon. C. (1949) *The mathematical theory of communication.* University of Illinois, Urbana.
Shulman, A. R. (1970) *Optical data processing.* Wiley, New York.
Smith, H. M. (1975) *Principles of holography,* 2nd edn. Wiley, New York.
Smith, F. G. (1966) *Radio astronomy,* 3rd edn. Penguin Books, London.
Smith, F. G. and Thomson, J. H. (1971) *Optics.* Wiley, London.
Stark, H. (ed.) (1982) *Applications of optical Fourier transforms.* Academic Press, New York.
Stokes, A. R. (1952) Three-dimensional diffraction theory of microscope image formation. *Proc. Roy. Soc. A,* **212**, 264–274.
Stone, J. M. (1963) *Radiation and optics.* McGraw-Hill, New York.
Stroke, G. W. (1969) *An introduction to coherent optics and holography,* 2nd edn. Academic Press, New York.
Taylor, C. A. (1978) *Images.* Wykeham Publications, London.
Taylor, C. A. and Lipson, H. (1964) *Optical transforms.* Bell, London.
Welford, W. T. (1981) *Optics,* 2nd edn. OUP, Oxford.
Wiener, N. (1933) *The Fourier integral and certain of its applications.* CUP, London.
Wiener, N. (1949) *The extrapolation, interpolation and smoothing of stationary time series.* Technology Press and Wiley, New York.
Williams, E. W. (1941) *Applications of interferometry.* Methuen, London.

Index

A

Abbe, E., 84, 90
Abbe principle, 93
Abbe (-Porter) theory of image formation, 84, 90-
aberrations
   correction (compensation), 112, 115
   and transfer functions, 89
'aether', 127
Airy, Sir George, 25, 32
Airy disc, 32
Airy pattern, 25, 32, 39
   amplitude, 39
   intensity, 39
   in image formation, 84, 100
alphanumeric pattern recognition, 116
   see also character recognition, pattern recognition.
amplified sampling of sinc function, 38
amplitude filter, 107
amplitude object, 110
amplitude transmission factors ('propagators'), 134
amplitude transmittance, 31
angular diameter of a star
   definition, 19, 33, 120
   measurement, 19, 120, 154
angular resolution limit, *see* resolution limit
aperture function
   as optical 'structure', 32
   of single aperture, 33
   of grating, 51
aperture mask, 12
aperture stop, 36
aperture synthesis, 146, 151
apodization
   in correcting aberrations, 34
   in Fourier transform spectroscopy, 143
Appleton, E. V., 146
astronomical telescope, 33

astronomy (optical and radio), 145-
autocorrelation function
   complex autocorrelation function $\Gamma_{11}(\tau)$, 135, 138
   see also coherence
   definition, 79
   Fourier transform of (Wiener-Khinchin theorem), 81

B

bacteriophage, 108
bandwidth, 76
   theorem, 77
baseline, 121, 147-
beam expander, 19, 94
beamsplitter, 128, 140, 144
Berger, J. E., 107
Bessel function, 32
Betelgeuse, 122
blocking filter, 107
blurring function, 72
Blythe, J. H., 151
Bolton, J., 147
de Broglie wavelength of electron, 102
Bragg, W. H., 44
Bragg, W. L., 44, 96-
Bragg equation, 164-
Bragg reflection, 164-

C

camera, 35
carrier fringes, 105
Cassiopeia, 148, 154
character recognition (pattern recognition), 114-
circular aperture
   aperture function, 33
   diffraction pattern, 32
van Cittert-Zernike theorem, 138

Index

coherence, 16-
  and source size, 17
  and correlation, 133-
  and spectral distribution, 16
  and visibility, 16, 133-
  area, 19
  complex degree of mutual coherence (or cross-correlation) $\gamma_{12}(\tau)$, 135
    (*also* the 'normalized complex mutual coherence function' *and* 'phase coherence factor')
  complex mutual coherence (or cross-correlation) function $\Gamma_{12}(\tau)$, 135
  complex self-coherence (or autocorrelation) function $\Gamma_{11}(\tau)$, 135, 139
  degree of mutual coherence $|\gamma_{12}(\tau)|$, 136
    (*also* the partial coherence)
  degree of self-coherence $|\gamma_{11}(\tau)|$, 139
  length, 17, 77
  mutual coherence, $[\Gamma_{12}(\tau)]$, 134
  partial coherence (*also* degree of mutual coherence) $|\gamma_{12}(\tau)|$, 16, 133, 136
  phase coherence factor $\gamma_{12}(\tau)$, 135
    (*also* complex degree of mutual coherence)
  spatial, 16-
  temporal (time), 16-, 75
  transverse, 19
  width, 19
coherent imaging, 20, 90
  of non-periodic objects, 93
  of periodic objects, 90
coherent optical processing *see* optical processing (coherent)
complex amplitude, 161
composition product, 72
conjugate parameters, 64 (x,u), 75 (t,v)
conjugate planes, 25, 94
contrast, 15
convolution ⊛, 70-
  and correlation, compared, 80
  and image formation, 84
  and multiplication, 114
  definition, 72
  image as, 115
  theorem, 74
  with array of δ-functions, 73
correlation ⊙, 77-
  and coherence, 135
  and convolution, compared, 80
  and interferometry, 148, 155
  and visibility, 133-
  definition, 77-
  see also coherence
correlation interferometer, 155
cos² fringes, 22, 120, 126, 129
cosine Fourier transform, 63, 126, 141

Crab Nebula, 147
cross-correlation
  complex cross-correlation function $\Gamma_{12}(\tau)$, 135
  definition, 79
  function, 79
  image, 114
  see also coherence
crystal
  as a 3-dimensional grating, 44-
  diffraction, 44-, 55-, 164-
  lattice, 44
  structure determination, 45, 96-
crystallography
  X-ray, 45, 96-
Cygnus, 146, 154

D

deblurring, 115
degree of mutual coherence $|\gamma_{12}(\tau)|$, 136
  see also coherence
degree of self-coherence $|\gamma_{11}(\tau)|$, 139
  see also coherence
delta function (δ-function), 68
  array, 69
  convolution, 73
  definition, 68
  Fourier transform of, 69
diffraction, 12
  and interference, 13
  and scattering, 20, 44
  -based optical processing, 116
  far field, 23
  Fraunhofer, 20-
  Fresnel, 23
  near field, 23
  scalar-wave treatment, 11, 157-
  theory of image formation, 90-
  X-ray, 44, 164-
diffraction-limited lens, 34
diffraction gratings
  1-dimensional, 39
  2-dimensional, 43
  3-dimensional, 44
diffraction order, 37, 41
diffractometer, optical, 94
digital imaging and processing, 83, 108
diopside, 98
double-aperture diffraction pattern, 36-
  as sampling of single-aperture pattern, 39
Duffieux, P. M., 85
  on image formation, 85
  on transfer functions, 85

E

electrical networks, 85
electromagnetic spectrum, 169

electron density distribution in crystals, 55–
electron microscopy, 100, 102, 107–
    and holography, 102
    resolution limit, 102
    spatial filtering, 106
electron wavelength, 102
energy spectrum correlator, 116
    see also power spectrum
entrance pupil, 36
Elias, P., 85
eye, as an imaging system, 90
eyepiece, role of, 35
exit pupil, 36
exit pupil function, 36

F

Fabry–Perot interferometer, 133
faltung (German), 72
    see also convolution
far-field approximation, 24
far-field diffraction, 23
Fellgett, P., 145
Feynman, R., 170
filter
    amplitude, 107
    blocking, 107
    complex, 107, 112
    holographic, 112
    low-pass, 106
    matched, 115
    phase, 110
    spatial frequency, 100, 106
    Vander Lugt, 114
Fizeau, 121, 130
flux, 13
folding product, 72
    see also convolution
Fourier, J. B. J., 11
Fourier analysis, 51–
Fourier aspects of image formation, 90–
Fourier–Bessel transform, 148
Fourier integral, 64
Fourier plane, 55, 114
Fourier series, 49, 50–
    and image formation, 92
    and diffraction gratings, 54–
    coefficients, 51, 52
    exponential notation, 58
    harmonics, 51
    sine and cosine series, 57
Fourier space, 55, 67
Fourier synthesis, 51
Fourier theorem, 49
Fourier transform, 61–
    and diffraction by apertures and gratings, 61–, 67–, 93

Fourier transform – *continued*
    and light waves, 75–
    cosine, 63
    exponential notation, 63
    hologram, 113, 114
    of δ-functions, 69
        rectangle function, 66
        sinc function, 66
    pairs, 64, 66, 75
Fourier transform spectroscopy, 139–
    basic equation of, 141
Fourier transformation performed by lens, 67, 94
Fourier transformation, relationships in interferometry, 125, 132, 138
Fraunhofer diffraction pattern, 20–, 28–
    as Fourier transform of aperture function, 66
    circular aperture, 32
    double-slit, 36
    N-slit grating, 39
    single-slit, 28
    two circular apertures, 36
Fraunhofer hologram, 113
frequency
    domain, 55, 106
    response, 88
    space, 55, 67
    spatial, 52
    spatial filtering, 106
    temporal, 75
Fresnel, 12
    diffraction pattern, 13, 23
    hologram, 113
Friedel's law, 48
fringes
    $\cos^2$, 22
    Young's, 12
fringe visibility
    and brightness distribution, 123–
        spectral distribution, 126, 131–
    definition, 15
    see also visibility of fringes

G

Gabor, D., 100, 102
general interference law for partially coherent light, 135
general interference law for stationary optical fields, 135
generalized hologram, 113
geometrical optical transfer function, 88
geometrical optics, 12
    and imaging, 90
    -based processing, 117

Index

grating
  aperture function, 51
  as a convolution, 73–
  image formation of grating as object 90–
  repeat, 40
  term, 40
grating diffraction pattern
  as a product of transforms, 67–
  as a sampling of single-slit pattern, 38, 42
  diffraction order 41
  double aperture 36–
  N-slit, 39–
  principal maxima, 40
  secondary (subsidiary) maxima, 41
  3-dimensional, 44
  2-dimensional, 43

H

Hanbury Brown, R., 153–
harmonics, 51
helium–neon laser *see* laser
Hey, J. S., 146
holographic filter, 112
holography, 102–
hologram, 102–
  Fourier transform, 113
  generalized, 113
  Leith–Upatnieks, 104
  See also Gabor, D.
human wart virus, 109
Huygens
  principle, 12
  secondary wavelets, 12

I

illumination intensity, 13
illuminance, 13
image
  as a Fourier synthesis, 92
  as a Fourier transform, 93
image contrast control, 110
image formation, 19–
  Abbe (–Porter) theory, 84–
  as a double Fourier process, 92
  as a double process of diffraction, 92
  convolution aspects of, 84, 86–, 100
  Duffieux on, 85
  electron microscope, 102
  Fourier aspects of, 90–
  geometrical optics v. diffraction theory, 90–
  importance of phase data, 109
  in coherent light, 90–
  in incoherent light, 86–

image formation – *continued*
  non-periodic objects, 93–
  periodic objects, 90–
  Rayleigh theory, 84, 100
impulse, 34
  function, 68
  response, 34
inclination factor, 14
incoherent illumination, 16–
incoherent imaging, 86–
incoherent optical processing *see* optical processing (incoherent)
information retrieval, 116
information theory, 86
infrared, 169
  Fourier transform spectroscopy, 139–
intensity interferometer, 153–
intensity (power) spectrum, 82, 138
  see also power spectrum
interference, 13
  and diffraction (terminology), 13
  by division of amplitude, 13, 25–, 127
  by division of wavefront, 13
  $\cos^2$ fringes, 22
  microscopy, 112
interferogram, 141
interferometer
  baseline, 121
  correlation, 154–
  intensity, 153–
  radio, 146–
  spectral, 127–
  stellar, 120–
interferometry, 119–
  and correlation, 148, 154–
invariance, 72, 86
invariant linear system, 86
inverse square law, 14
inverse transform, 64
ion cyclotron resonance spectroscopy (ICR), 145
irradiance, 13
isoplanatic system, 87

J

Jacquinot, P., 145
Jansky, K. G., 146
Jennison, R. C., 153
Jodrell Bank, 153–
Jupiter, 121

K

Kirchhoff, 12, 14
Klug, A., 100, 107–

L

language translation, 116
laser
   coherence, 16, 77
   helium–neon, 94
   in holography, 103–
   in optical diffractometry, 94
   light, 16, 77
lattice
   translation vectors, 45
   see also crystal lattice
Laue, Max von, 44
Laue conditions (equations), 45, 164
Laue experiment, 44
Leith, E. N., 103
Leith–Upatnieks holography, 104–
lens
   aberrations, 100
   aperture, 25, 36, 100
   as a Fourier transformer, 67, 93
   diffraction- limited, 34
   transfer functions as assessment of, 89
light
   monochromatic, 16
   quasimonochromatic, 16
   spectrum, 169
   thermal, 16, 77
   white, 16, 77
   see also laser
limit of resolution, see resolution limit
linear system, 72, 85
line-spread function (LSF), 72, 89
Lipson, H., 94–, 100
Lloyd's mirror, 146
Lovell, A. C. B., 154
low-pass filter, 106

M

mask, 12
   trial mask, 97
mass spectroscopy, 145
matched filter, 115
Michelson, A. A., 119–
Michelson spectral interferometer, 127–, 139
   fringe visibility and spectral distribution, 131
   wavelength measurement, 127
Michelson stellar interferometer, 19, 120–
   fringe visibility and brightness distribution, 123–
   measurement of stellar diameters, 120–
Michelson–Morley experiment, 127
microscope see image formation
microscope objective
   numerical aperture, 36
   oil-immersion, 35

microscopic limit, 36
Mills, B. Y., 150
Mills cross, 150
monochromatic light, 16
modulation, 15, 87
modulation transfer factor, 88
modulation transfer function (MTF), 88
Morley, E. W., 127
Mullard Radio Astronomy Observatory, 151
multiple-beam interference, 133
multiplexing, 145
mutual coherence, $\mathcal{R}[\Gamma_{12}(\tau)]$, 135
   see also under coherence

N

near-field diffraction, 23
Newton's rings, 13, 25, 130
Nomarski differential-interference-contrast microscopy, 112
non-crystalline materials, structure of, 46
non-periodic objects
   imaging of, 93
nuclear magnetic resonance spectroscopy (NMR), 145

O

object
   amplitude, 110
   phase, 110
obliquity factor, 14
optical astronomy, 153–
optical diffractometer, 94
optical filtering, 106
   see also optical processing
optical imaging, 83–
   see also image formation
optical masks, 94–
optical multiplier and integrator, 117
optical path difference, 26
optical processing (coherent), 106–
   aberration compensation, 115
   amplitude filter, 107
   applications, 106–
   blocking filter, 107
   character recognition (pattern recognition), 114
   complex filter, 107, 112–
   convolution image, 115
   cross-correlation image, 114
   deblurring, 115
   differential-interference-contrast microscopy, 112
   electron microscopy, 107–
   filtering, 106–
   holographic filter, 112

Index

optical processing (coherent) – *continued*
  interference microscopy, 112
  low-pass filter, 106
  matched filter, 115
  Nomarski microscope, 112
  optical set-up for processing, 106
  pattern recognition (character recognition), 114
  phase-contrast (Zernike) microscope, 110
  phase filter, 111
  Vander Lugt filter, 114
optical processing (incoherent), 115–
  alphanumeric pattern recognition, 116–
  diffraction-based, 116
  energy spectrum correlator, 116
  geometrical optics-based, 117
  information retrieval, 116
  language translation, 116
  optical multiplier and integrator, 117
  pattern (character) recognition, 116–
optical structure and aperture function, 32
optical transfer function, 85
  as autocorrelation of aperture function, 89
  as Fourier transform of point-spread function (PSF), 87–
  determination of, 88–
  geometrical, 88
  modulation, 88
  phase, 88
  see also transfer function
optical transforms, 94–
order of diffraction, 37, 41
parallel plate, 26
partial coherence, 16, 133–
  see also degree of mutual coherence
path difference, 159
pattern recognition (character recognition), 114, 116–
Pawsey, J. L., 146
Pease, F. G., 123
phase, 158
  angle, 158
  change on reflection, 163
  closure, 153
  coherence factor $\gamma_{12}(\tau)$, 135
  constant, 158
  difference, 159
  filter, 110
  object, 110
  plate, 111
  problem of X-ray crystallography, 97, 100
  transfer function (PTF), 88
phase-contrast microscope, 110
phase switching, 149
phasor diagram, 160
photon, 16, 76

phthalocyanine, 97
plane-wave approximation, 24
point-spread function (PSF), 34, 72, 87–
polarization of light, 157
Porter, A. B., 84, 90
  see also Abbe–Porter
positron emission tomography, 110
power spectrum
  spatial, $|F(u)|^2$, 82
  temporal, $|F(v)|^2$, 82, 139
principal maxima, 41
principle of reversibility, 92
principle of superposition, 13, 159
propagators, 134
pseudoscopic image, 105
pulsar, 153
pulse, definition, 77
pupil
  entrance, 36
  exit, 36
  function, 36

Q

quasimonochromatic light, 16

R

radiation sources
  structures of, 119–
radio waves
  coherence of, 19
radio interferometry, 146–
radio spectrum, 169
radioastronomy, 145–
ray optics, ray theory, 12, 90
  see also geometrical optics
Rayleigh, Lord (J. W. Strutt), 84
Rayleigh
  criterion for resolution, 34–
  on image formation, 84–, 100
  on visibility curves, 126, 132
reciprocal lattice, 44, 96
reciprocal space, 31, 55
reconstruction of wavefront, 102–
rectangle function, 32, 62
  Fourier transform, 66
reflection, phase change on, 163
resolution criterion *see* Rayleigh
resolution limit, 93
  Fourier transform spectrometer, 143
  microscope, 35
  telescope, 34
  stellar interferometer, 122
  spectral interferometer, 130
Rogers, A. E., 153
Ryle, Sir Martin, 147–, 153

S

sampled sinc function, 39
scalar-wave model of light, 157–
scalar-wave theory of diffraction, 12–
scattering and diffraction (terminology), 20, 44
Schade, O. H., 85
Schwarz inequality, 135
secondary (subsidiary) maxima, 41
secondary wavelets, 12
self-coherence *see* coherence
self-luminous object, 20, 35
Shannon, C., 86
sinc function, 31
    sampled, 39, 42
single-slit aperture function, 32
single-slit diffraction, 28–
Sirius, 155
Smith, F. G., 148
smoothing function, 72
source
    monochromatic, 16
    quasimonochromatic, 16
    size and coherence, 17–
    thermal, 16
space
    Fourier, 55, 67
    frequency, 55, 67
    reciprocal, 31, 55
space telescope, 155
spatial coherence, 16–, 137
spatial filtering, 100, 106–
spatial frequency, 52
    plane (domain), 55
spatial invariance, 86
spectral interferometer *see* Michelson spectral interferometer
spectral linewidth, 76
spectroscopy
    Fourier transform, 139–
    mass, 145
    ion cyclotron resonance (ICR), 145
    nuclear magnetic resonance (NMR), 145
squamous cells, 112
Stainer, H. M., 148
Stanley, G., 147
star
    angular diameter, 19
    image as Airy pattern of telescope objective, 33
stationarity, 86
stationary system, 134
statistical network theory, 86
stellar interferometer *see* Michelson stellar interferometer
Stokes' treatment of reflection and refraction, 163
stop, aperture, 36
subsidiary (secondary) maxima, 41
Sun, 147–
superposition, principle of, 159
superposition integral, 72
supersynthesis, 151–
system
    definition, 85
    frequency response of, 88
    isoplanatic, 87
    linear, 85
    space-invariant, 86
    time-invariant, 86
    (unit) impulse response of, 34, 86

T

Taurus, 147
Taylor, C. A., 100
telescope
    astronomical, 33
    radio, 145–
    resolution limit, 34, 146
temporal coherence, 16, 76, 138–
temporal frequencies, 75
    and Fourier transform, 75
temporal invariance *see* invariance
thermal light, 77
3-dimensional grating, 44
time-invariance, 86
top-hat function, 32
transfer function, 85
    as unit impulse response, 85
    of linear system, 85
    see also optical transfer function
transform, *see* Fourier transform
transverse coherence, 19
Twiss, R. Q., 155
2-dimensional grating, 43, 94–
2-wavelength microscope, 99, 102
Twyman–Green interferometer, 140

U

unit impulse function, 68
unit impulse response, 85
ultraviolet, 169
    Fourier transform spectroscopy, 145
Uptanieks, J., 103
    see also Leith–Upatnieks
unfilled aperture, 150

V

van Cittert–Zernike theorem, 138
Vander Lugt, A.
    filter, 114

Index

Venus, 144
visibility, definition, *see* visibility of fringes
visibility curve (visibility function) $V(D)$, 125–, 129, 141–
visibility of fringes
  and coherence, 137–
  and correlation, 137–
  definition, 15
visibility function, $V(D)$, 125
visible spectrum, 169
Vonberg, D. D., 147–

W

wave group (packet), 76
wave optics, 90
wavefront reconstruction, 102–
wavenumber $(1/\lambda)$, 158
waves
  addition by phasor diagram, 160
  use of exponential notation, 161
wavetrain, 16–
white light, 16, 77
Wiener, N., 86
Wiener–Khinchin theorem, 81, 139, 143

X

X-ray
  analysis of crystal structure, 44, 96–
  computerized tomography (X-ray scanner), 110
  diffraction, 44
  microscope, 99
  reflection, 46–, 55, 164
  spectrum, 169
X-ray crystallography, 44, 96–
  phase problem in, 97, 100

Y

Young, Thomas, 11
Young's experiment, 12
  fringes, 12

Z

Zeiss, C., 90
Zernike, F.
  and holography, 102
  and image formation, 91–
  and phase-contrast microscopy, 110–
  see also van Cittert–Zernike
zeugmatography, 110